Notable Numbers

William T. Stokes

Covington Middle School
Los Altos School District
Los Altos, California

The author and publisher hereby grant permission to reproduce any part of this book for use in the classroom.

Dedicated to my wife Evelyn ——

for sharing in the many hours devoted to this book.

Drawings by the author

Cover design by Judy Ammon

Standard Book Number 0-914534-01-7

Post Office Box 415, Palo Alto, California 94302

Introduction

For many centuries, early mathematicians found that a sensitivity to number patterns helped them to solve difficult problems. If the mathematician found an unusual pattern or sequence of numbers, he then sought a rule or formula that would explain or describe the pattern. The discovery of such unique number patterns proves to be stimulating and enjoyable, not only to mathematicians, but to students at every grade level.

Over the years, my students have been fascinated by number relationships. This study is particularly motivational to students normally not interested in any kind of mathematics. There is an apparent need for enrichment material not found in the typical classroom text. Consequently, many mathematics teachers have found themselves in a classroom situation attempting to help students visualize a pattern or concept, but lacked the time and technical skill needed to construct the necessary visuals. All too frequently, such makeshift diagrams failed to meet the graphic requirements or proved ineffective due to disorganized additions and revisions. This book presents a collection of successful lessons, worksheets, and transparency visuals for the immediate use of the classroom teacher.

Most of the topics presented are intended to serve as introductions to the further exploration of particular topics. A list of references has been provided in the back of this book to assist the teacher in finding additional information about the major topics.

The author and publisher hope these materials prove valuable to you; we welcome your reactions and any suggestions for additional topics you would like to see in later editions of this book.

How to use this book

ORGANIZATION OF MATERIAL

After selecting a lesson, use the original to produce a transparency and thermal spirit master. As a teacher who accumulates a library of such resource materials, you will find it necessary to organize them in an efficient filing system. Such a system would provide the teacher and students with an instant inventory of available lessons.

To help establish a file, this book features unique title tabs lettered in large bold type. The title tabs may be cut from the margin of the transparency, an area which is ordinarily covered by the transparency frame. Although the title tabs have been designed to fit the two inch tabs which accompany Pendaflex file folders, they can also be scotch taped to manila folders with similar success. Pendaflex folders are recommended for easy access and durability.

Colored paper may be used to back the title tabs to allow for quick identification of related topics.

THERMAL SPIRIT MASTERS

The type size and page layout in this book were specifically designed to create the best possible clarity for student worksheets and overhead transparencies. To produce the sharpest impressions use a carrier device, such as the Old Town Copymaker Reflecto-Shield. The blue copy is non-reproducible in nearly every type of thermal-master machine. Follow the instructions given with the particular brand of thermal masters used.

OVERHEAD PROJECTION TRANSPARENCIES

To obtain best results, use a heavy weight transparency film which will produce a BLACK image on a positive clear background. 3M Brand type 127 film or its equivalent may be used successfully.

First adjust the settings of the thermal transparency maker to the manufacturer's specifications. Then place the transparency film upon the original master so that it is centered. Finally, feed the film and master into the machine together. The outcome will be a transparency which is ready to mount in a standard (7½″ x 9½″) cardboard frame.

If additions or deletions are desired, they should be made directly on the original master. Information can be added by typing with a black carbon ribbon or writing with a soft (No. 2) pencil. Copy can be deleted by covering the area with typewriter white out or liquid paper.

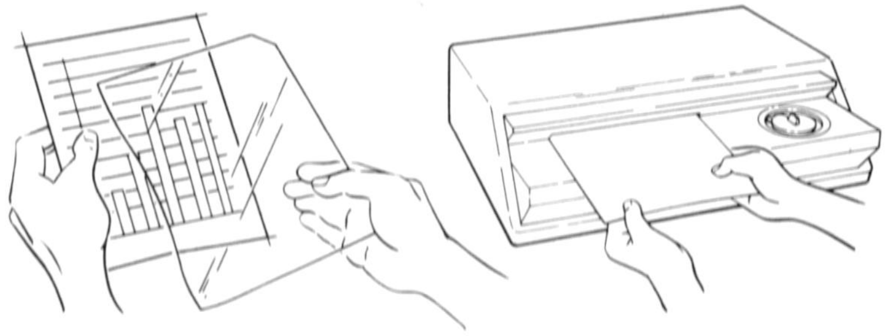

To add color, felt pens can be applied to the backside of the transparency. Use water-base inks to avoid any damage to the plastic. For even better results, try 3M Brand type 720 colored adhesive film.

ACTIVITY CARDS

Duplicated pages of this book lend themselves to use as activity cards. By simply inserting a duplicated page into an 8½ x 11″ plastic sheet protector, a teacher can make a reusable activity card. Students should write on these using felt tip pens containing water soluable ink so that their work can then be removed with a damp cloth or tissue. For easy access, the mounted lessons can be stored in a standard metal storage bin or in a three-ring notebook binder.

TABLE OF CONTENTS

Classification of Numbers

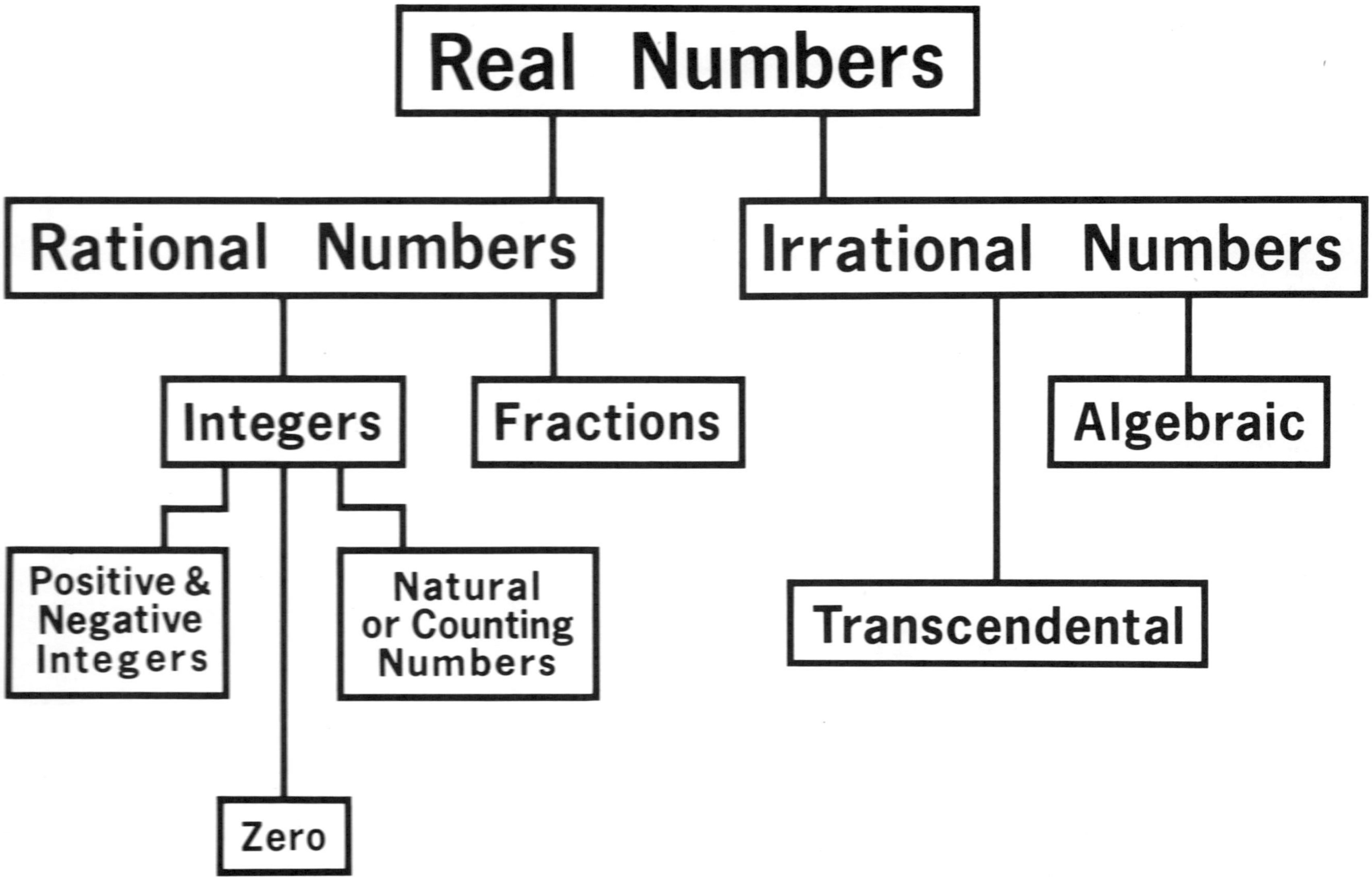

THE HISTORY OF NUMERALS

EGYPTIAN HIEROGLYPHIC NUMERALS

1	10	100	1,000	10, THS	100, THS	1 MILLION

STROKE	ARCH	COILED ROPE	LOTUS FLOWER	POINTED FINGER	TAD-POLE	ASTONISH-ED MAN

MODERN ROMAN NUMERALS

1	5	10	50	100	500	1,000
I	V	X	L	C	D	M

GENERATION OF HINDU-ARABIC NUMERALS

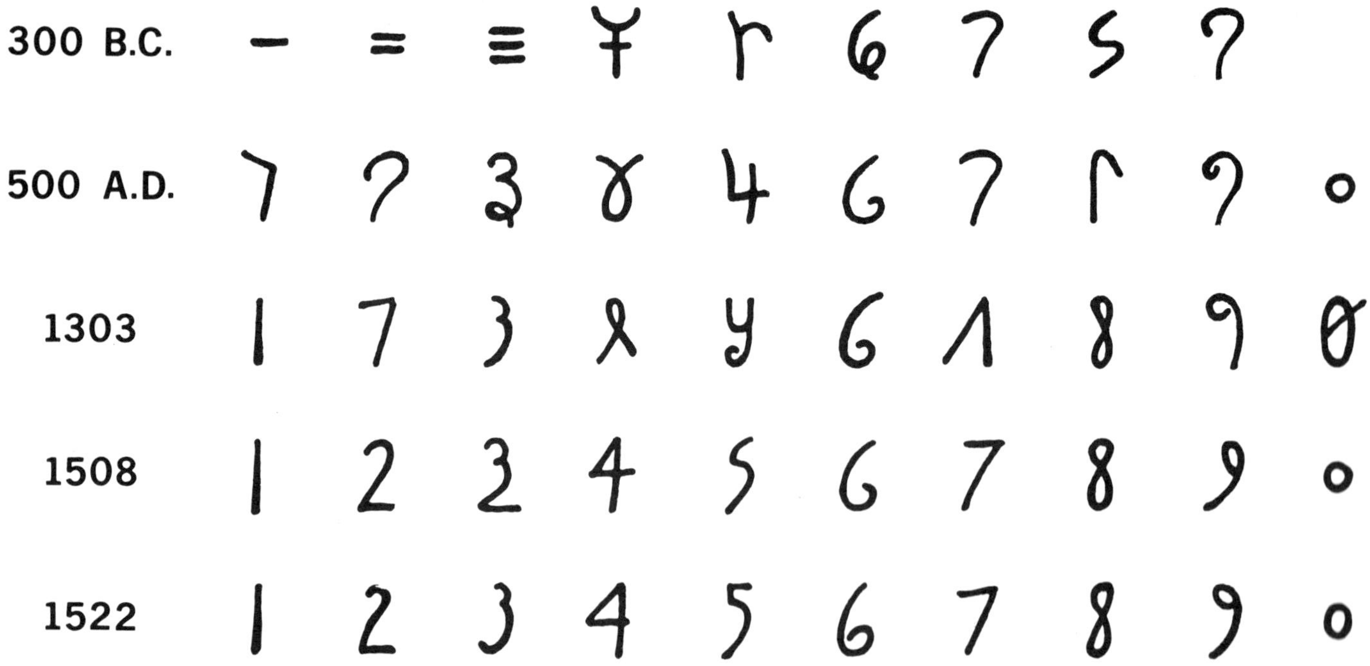

CARDINAL NUMBERS

A whole number which designates how many elements are contained in a set.

{A, B, C, D, E, F}

There are **6** letters

ORDINAL NUMBERS

A natural number that denotes the order of members of a set in a sequence or arrangement.

{A, B, C, D, E, F}

↑

'D' is the **4**th letter

AMICABLE NUMBERS

Two numbers are said to be amicable if each is equal to the sum of the proper divisors of the other.

Proper Divisors (of 220)

1
2
4
5
10
11
20
22
44
55
+ 110
= 284

Proper Divisors (of 284)

1
2
4
71
+ 142
= 220

Find the missing amicable numbers.

1,184 and 1,210.

17,296 and 18,416.

9,363,584 and 9,437,056.

All known amicable numbers are even numbers.

Proper Divisors —
All whole number divisors of a number except itself.

FACTORIAL NUMBERS

The product of a given series of consecutive whole numbers beginning with 1 and ending with the specified number.

$0! = 1$ (by agreement)

$1! = 1 \times 1 = 1$

$2! = 1 \times 2 = 2$

$3! = 1 \times 2 \times 3 = 6$

$4! = 1 \times 2 \times 3 \times 4 = 24$

$5! = 1 \times 2 \times 3 \times 4 \times 5 = 120$

$6! = 1 \times 2 \times 3 \times 4 \times 5 \times 6 =$ 720

$7! =$ 5,040 $9! =$ 362,880

$8! =$ 40,320 $10! =$ 3,628,800

Can you find the missing factorial numbers?

Factorial Numbers

FACTORIAL PHENOMENA

169

1! + 6! + 9!

Equals

1 + 720 + 362880

Equals

3	6	3	6	0	1

3! + 6! + 3! + 6! + 0! + 1!

Equals

6 + 720 + 6 + 720 + 1 + 1

Equals

1	4	5	4

1! + 4! + 5! + 4!

Equals

1 + 24 + 120 + 24

Equals

1	6	9

ABUNDANT NUMBERS

A number is said to be abundant if the sum of its divisors is greater than its double.

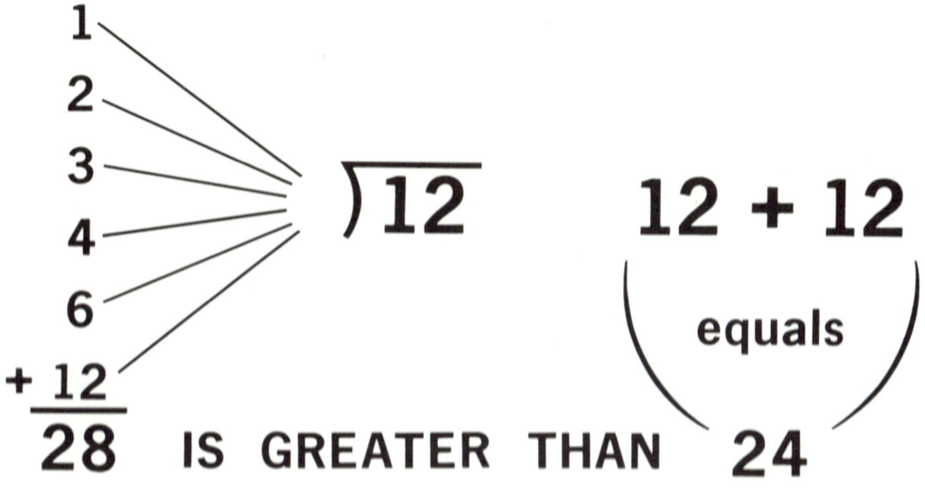

Circle the abundant numbers. Look for a pattern.

12	13	14	15	16
17	18	19	20	21
22	23	24	25	26
27	28	29	30	31
32	33	34	35	36

Extend the chart to see if the pattern continues.

PALINDROMIC NUMBERS

A natural number of two or more digits which are the same either read forwards or backwards.

33, 585, 9229, 74047, 658856.

Conjecture —

Every integer can be made a palindromic number. Add its reverse to the integer; if the sum is not palindromic, add the sum and its reverse . . ., eventually this will produce a palindrome.

Procedure —

895	integer
+ 598	reversal 1
1493	sum
+ 3941	reversal 2
5434	sum
+ 4345	reversal 3
9779	palindrome

Notice —
The palindrome of the integer 895 is the same palindrome for the integer 598.

Practice —

Integer	Rev.	Palindrome
49	2	484
180	3	747
381	4	13431
275	5	44044
182	6	45254
395	7	881188
589	8	1136311
365	11	8855 5588
276	15	8836886388
98	24	See below

8813200023188

PERFECT NUMBERS

A perfect number is any natural number which equals the sum of its proper divisors, that is, all whole number divisors except the number itself.

	Perfect Numbers		Proper Divisors
If:	6	=	1 + 2 + 3
k = 3 →	28	=	1 + 2 + 4 + 7 + 14

Prove the following numbers are perfect numbers.

k = 5 →	496	=	1 + 2 + 4 + 8 + 16 + 31 + 62 + 124 + 248.
k = 7 →	8,128	=	1 + 2 + 4 + 8 + 16 + 32 + 64 + 127 + 254 + 508 + 1016 + 2032 + 4064.

Euclid's Formula for Perfect Numbers

$$PN = 2^{K-1} \times (2^K - 1)$$

$$PN = 2^{2-1} \times (2^2 - 1)$$

$$PN = 2^1 \times (4 - 1)$$

$$PN = 2 \times 3$$

$$PN = 6$$

provided $(2^K - 1)$ is a prime number

k = 13 → 33,550,336

Find the next 4 perfect numbers using K = 3, 5, 7, 13.

All known perfect numbers are even numbers ending in 6 or 28.

FIBONACCI NUMBERS

An infinite series of natural numbers in which each term is the sum of the two preceding terms.

1, 1, 2, 3, 5, 8, 13, 21 . . .

$$1 + 1 = 2$$
$$1 + 2 = 3$$
$$2 + 3 = 5$$
$$3 + 5 = 8$$
$$5 + 8 = 13$$
$$8 + 13 = 21$$

This sequence was the result of a problem posed by Leonardo Fibonacci in the twelfth century. Among the many contributions to mathematics, Fibonacci explained the use of Hindu-Arabic numerals in the computation of numbers. His famous rabbit problem read as follows:

> *"How many pairs of rabbits can be produced each month from a single pair in a year if each adult pair gives birth to a new pair every month, each new pair reproduces from the second month on, and no rabbit dies?"*

FIBONACCI NUMBER PATTERNS

Fibonacci Notation

F_1	F_2	F_3	F_4	F_5	F_6	F_7	F_8	F_9	F_{10}
1,	1,	2,	3,	5,	8,	13,	21,	34,	55.

$1 + 1 + 2 = 4$

$1 + 1 + 2 + 3 = 7$

$1 + 1 + 2 + 3 + 5 = 12$

$1 + 1 + 2 + 3 + 5 + 8 = 20$

$1 + 1 + 2 + 3 + 5 + 8 + 13 = 33$

$1 + 1 + 2 + 3 + 5 + 8 + 13 + 21 = 54$

SUMMATION OF FIBONACCI NUMBERS

The sum of the first n Fibonacci numbers is one less than the (n+2)nd Fibonacci number.

$$F_1 + F_2 + \ldots + F_N = F_{N+2} - 1$$

Fibonacci Numbers

1, 1, 2, 3, 5, 8, 13, 21, 34.

N + 2

−1

Summation

$1 + 1 + 2 + 3 + 5 + 8 = 20$

FIBONACCI NUMBER PATTERNS

Fibonacci Notation

F_1	F_2	F_3	F_4	F_5	F_6	F_7	F_8	F_9	F_{10}
1,	1,	2,	3,	5,	8,	13,	21,	34,	55.

Multiply

$1 \quad (2)^2 \quad 3$ (×)

$$4 - 3 = 1$$

$5 \quad (8)^2 \quad 13$

$$64 - 65 = -1$$

$21 \quad (34)^2 \quad 55$ (×)

$$1156 - 1155 = 1$$

$$(F_N)^2 - (F_{N-1})(F_{N+1}) = 1, -1, \text{ or } |1|$$

$$(8)^2 - (5 \times 13)$$

$$64 - 65 = -1 \text{ or } |1|$$

Can you discover from the pattern whether F_{22} will be −1 or 1?

F_1	F_2	F_3	F_4	F_5	F_6	F_7	F_8	F_9	F_{10}
1	1	2	3	5	8	13	21	34	55
	−1	1	−1	1	−1	1	−1	1	

All even F numbers -1

All odd F numbers 1

SQUARE NUMBERS

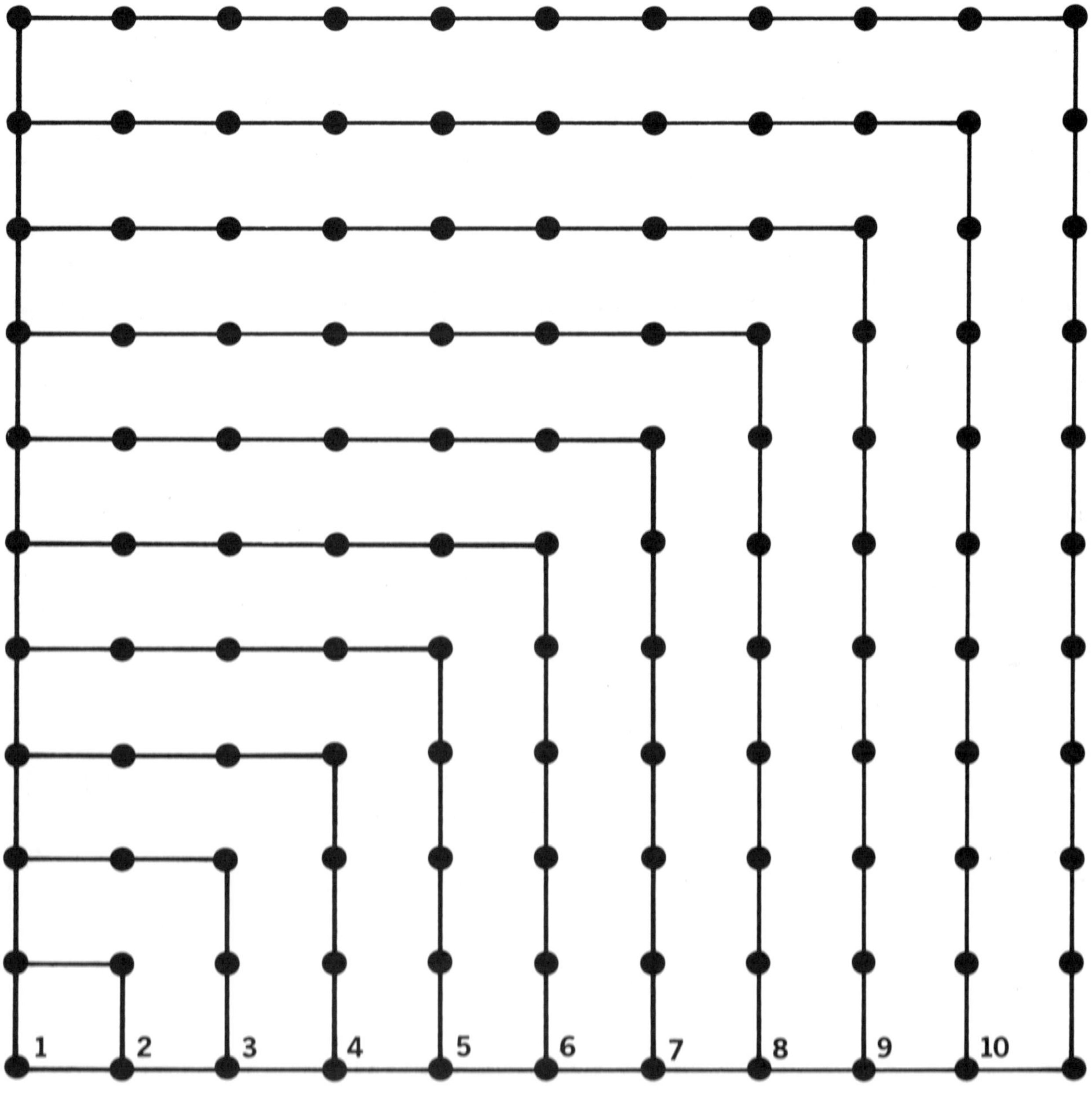

SQUARE NUMBERS

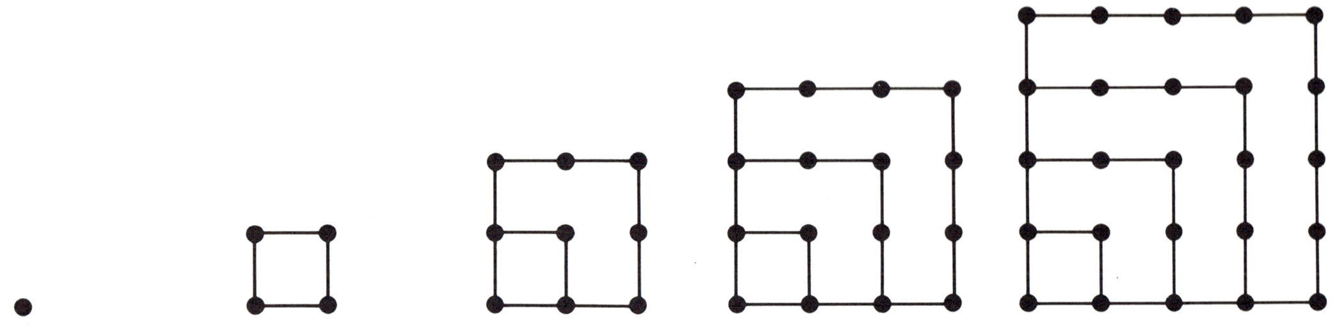

No. 1	No. 2	No. 3	No. 4	No. 5
$(1)^2 = 1$	$(2)^2 = 4$	$(3)^2 = 9$	$(4)^2 = 16$	$(5)^2 = 25$

No. 6	No. 7	No. 8	No. 9
$(6)^2 = 36$	$(7)^2 = 49$	$(8)^2 = 64$	$(9)^2 = 81$

No. 10	No. 11	No. 12	No. 13
$(10)^2 = 100$	$(11)^2 = 121$	$(12)^2 = 144$	$(13)^2 = 169$

Square Array

No. 14	No. 15	No. 16	No. 17	No. 18	No. 19	No. 20
196	225	256	289	324	361	400

TRIANGULAR NUMBERS

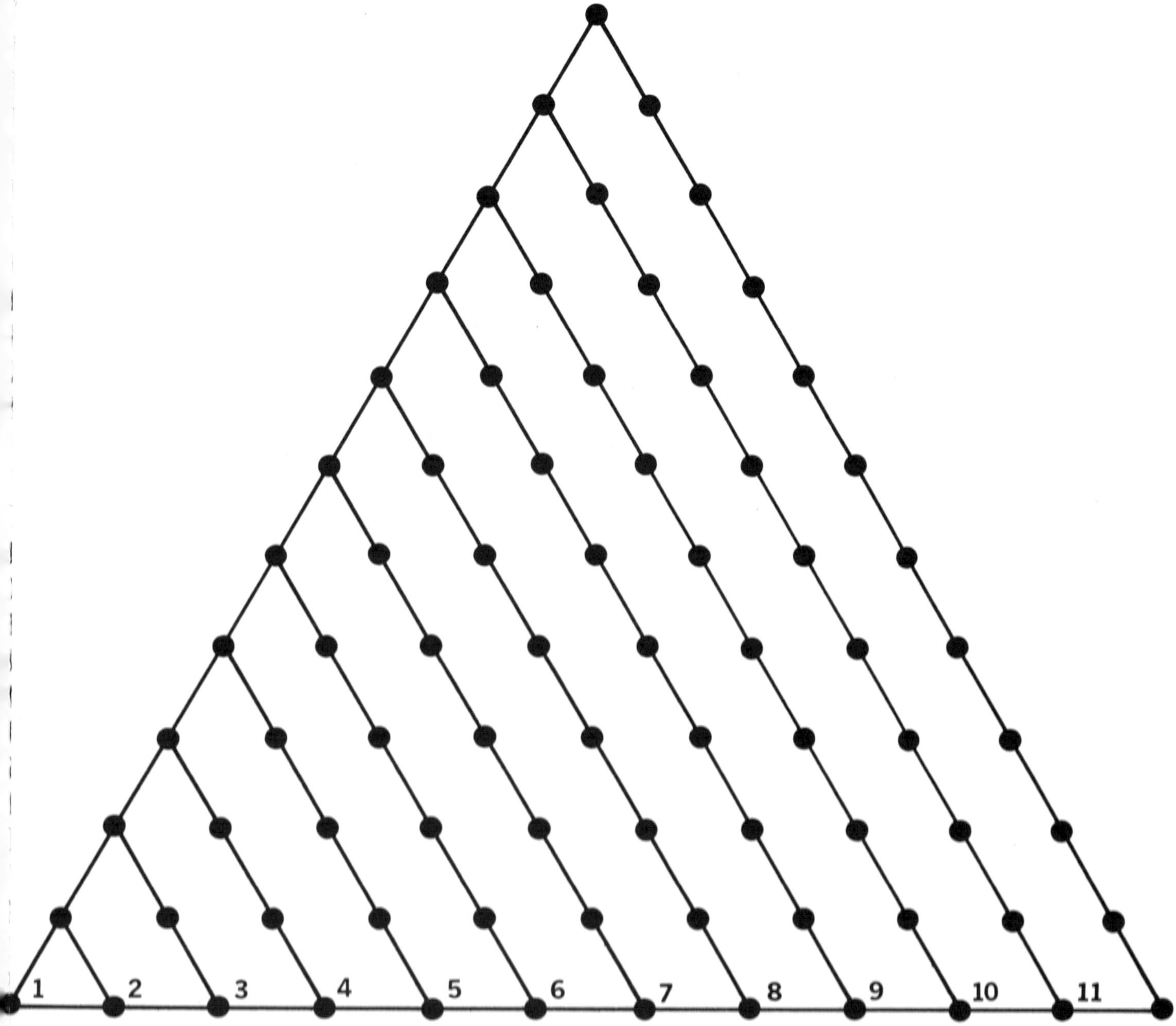

Triangular Number Array

TRIANGULAR NUMBERS

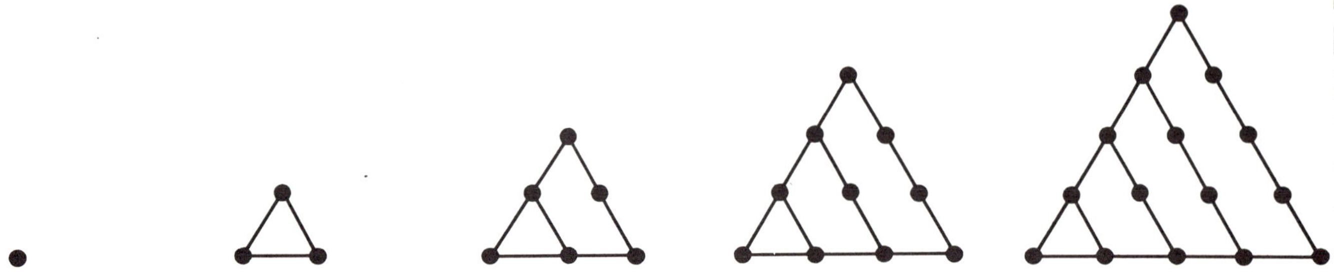

No. 1	No. 2	No. 3	No. 4	No. 5
$(1)^2 = 1$	$(2)^2 = 4$	$(3)^2 = 9$	$(4)^2 = 16$	$(5)^2 = 25$
1	4 – 1	9 – 3	16 – 6	25 – 10
1	3	6	10	15

No. 6	No. 7	No. 8	No. 9
$(6)^2 = 36$	$(7)^2 = 49$	$(8)^2 = 64$	$(9)^2 = 81$
36 – 15	49 – 21	64 – 28	81 – 36
21	28	36	45

Triangular Array

No. 10	No. 11	No. 12	No. 13	No. 14	No. 15	No. 16
55	66	78	91	105	120	136

The N th triangular number can be found by using the formula $\frac{N^2 + N}{2}$

Triangular Numbers

HEXAGON NUMBER PATTERNS

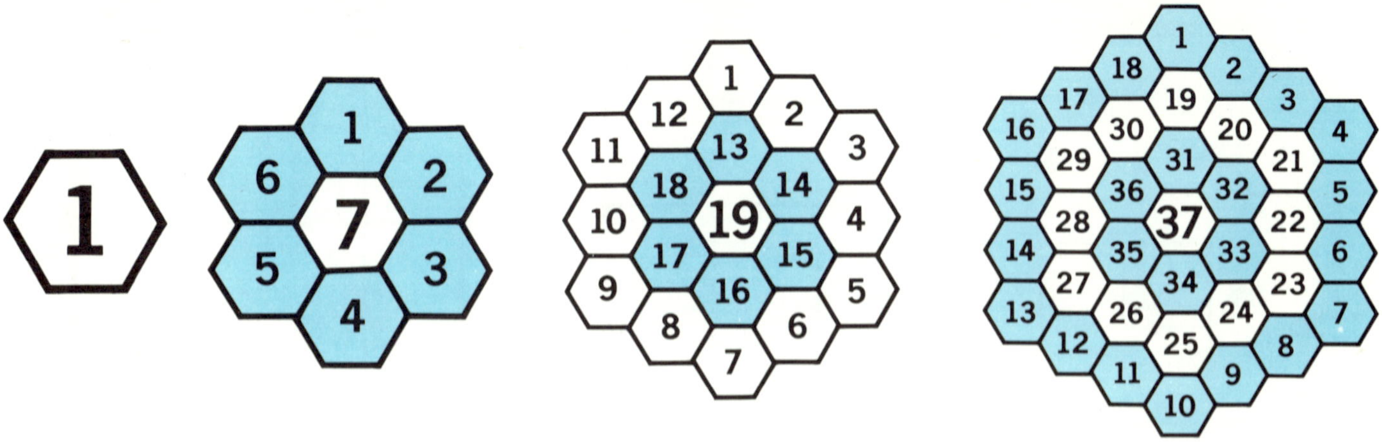

No. 1

$(1)^3 = 1$

1

1

No. 2

$(2)^3 = 8$

8 − 1

7

No. 3

$(3)^3 = 27$

27 − 8

19

No. 4

$(4)^3 = 64$

64 − 27

37

No. 5

$(5)^3 = 125$

125 − 64

61

No. 6

$(6)^3 = 216$

216 − 125

91

No. 7

$(7)^3 = 343$

343 − 216

127

No. 8

$(8)^3 = 512$

512 − 343

169

Hexagon Pattern

No. 9	No. 10	No. 11	No. 12	No. 13	No. 14	No. 15
217	271	331	397	469	547	631

The N th pattern can be found by using the formula $N^3 - (N-1)^3$

Hexagon Patterns

SUM OF SQUARES

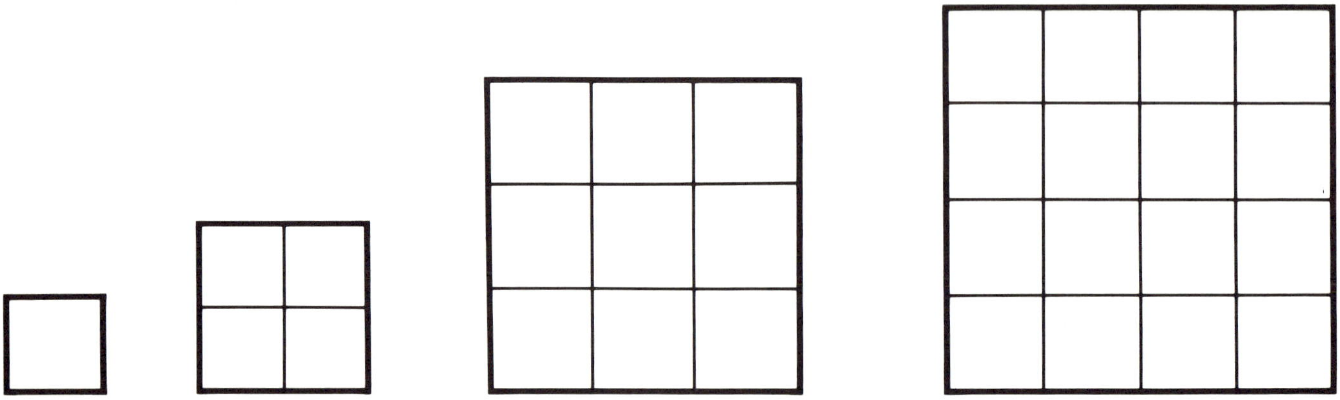

No. 1

$(1)^2$

1

1

No. 2

$(1)^2 + (2)^2$

1 + 4

5

No. 3

$(1)^2 + (2)^2 + (3)^2$

1 + 4 + 9

14

No. 4

$(1)^2 + (2)^2 + (3)^2 + (4)^2$

1 + 4 + 9 + 16

30

No. 5

$(1)^2 + (2)^2 + (3)^2 + (4)^2 + (5)^2$

1 + 4 + 9 + 16 + 25

55

No. 6

$(1)^2 + (2)^2 + (3)^2 + (4)^2 + (5)^2 + (6)^2$

1 + 4 + 9 + 16 + 25 + 36

91

Square Pattern

No. 7	No. 8	No. 9	No. 10	No. 11	No. 12	No. 13
140	204	285	385	506	650	819

The N th number of squares can be found by using the formula $\frac{N(2N+1)(N+1)}{6}$

PENTAGONAL NUMBERS

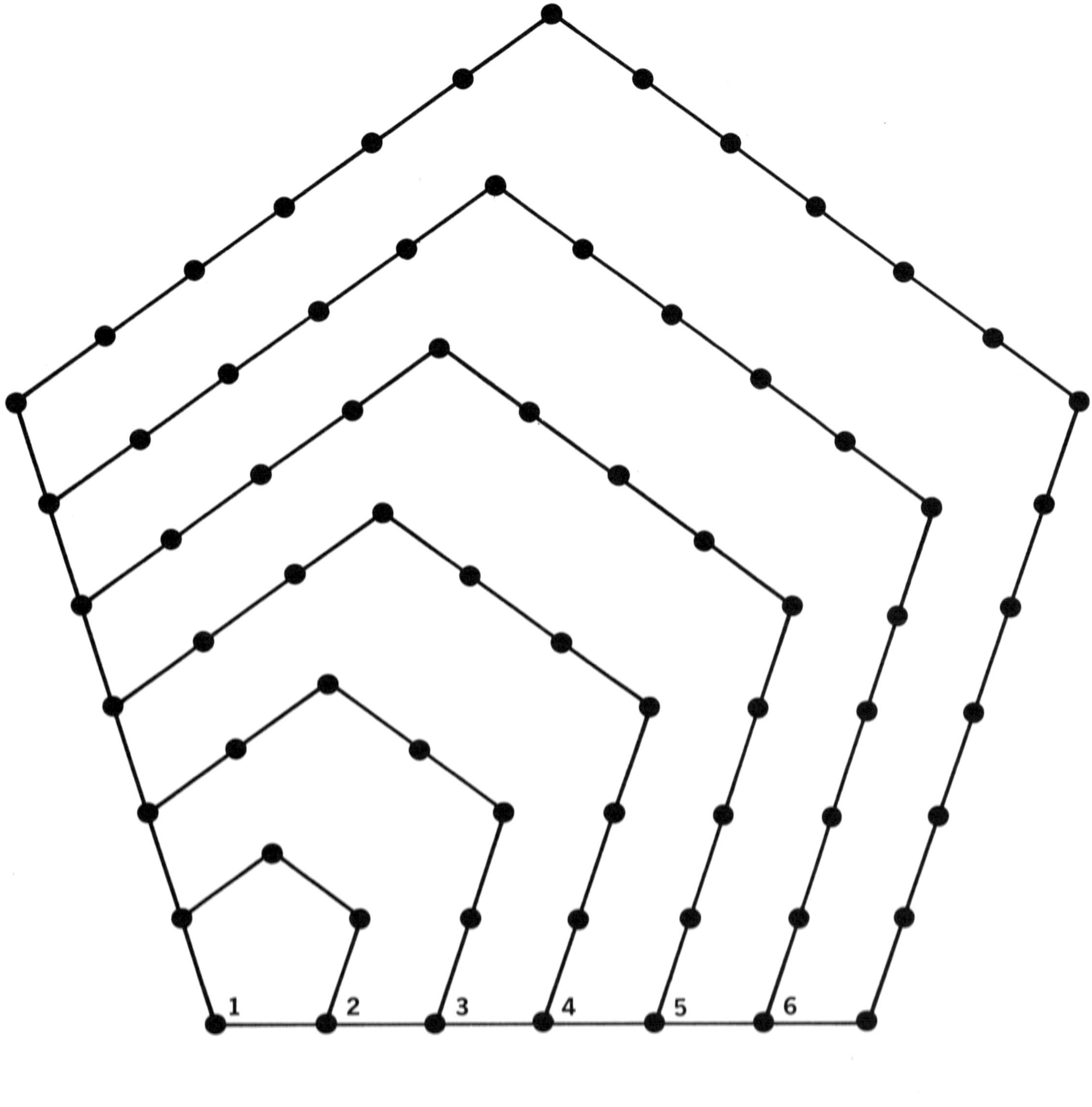

Pentagonal Number Array

PENTAGONAL NUMBERS

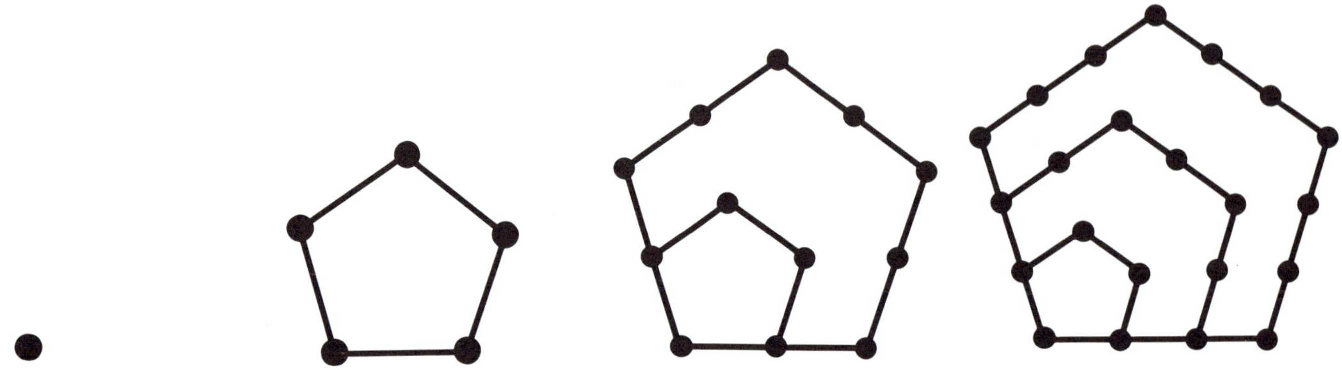

No. 1	No. 2	No. 3	No. 4
3(0) + 1	3(1) + 1	3(2) + 1	3(3) + 1
= 0 + 1	= 4 + 1	= 7 + 5	= 10 + 12
1	5	12	22
No. 5	No. 6	No. 7	No. 8
3(4) + 1	3(5) + 1	3(6) + 1	3(7) + 1
= 13 + 22	= 16 + 35	= 19 + 51	= 22 + 70
35	51	70	92

Pentagonal Array

No. 9	No. 10	No. 11	No. 12	No. 13	No. 14	No. 15
117	145	176	210	247	287	330

The Nth pentagonal number can be found by using the formula $\frac{N(3N-1)}{2}$

CIRCLE SECTIONS

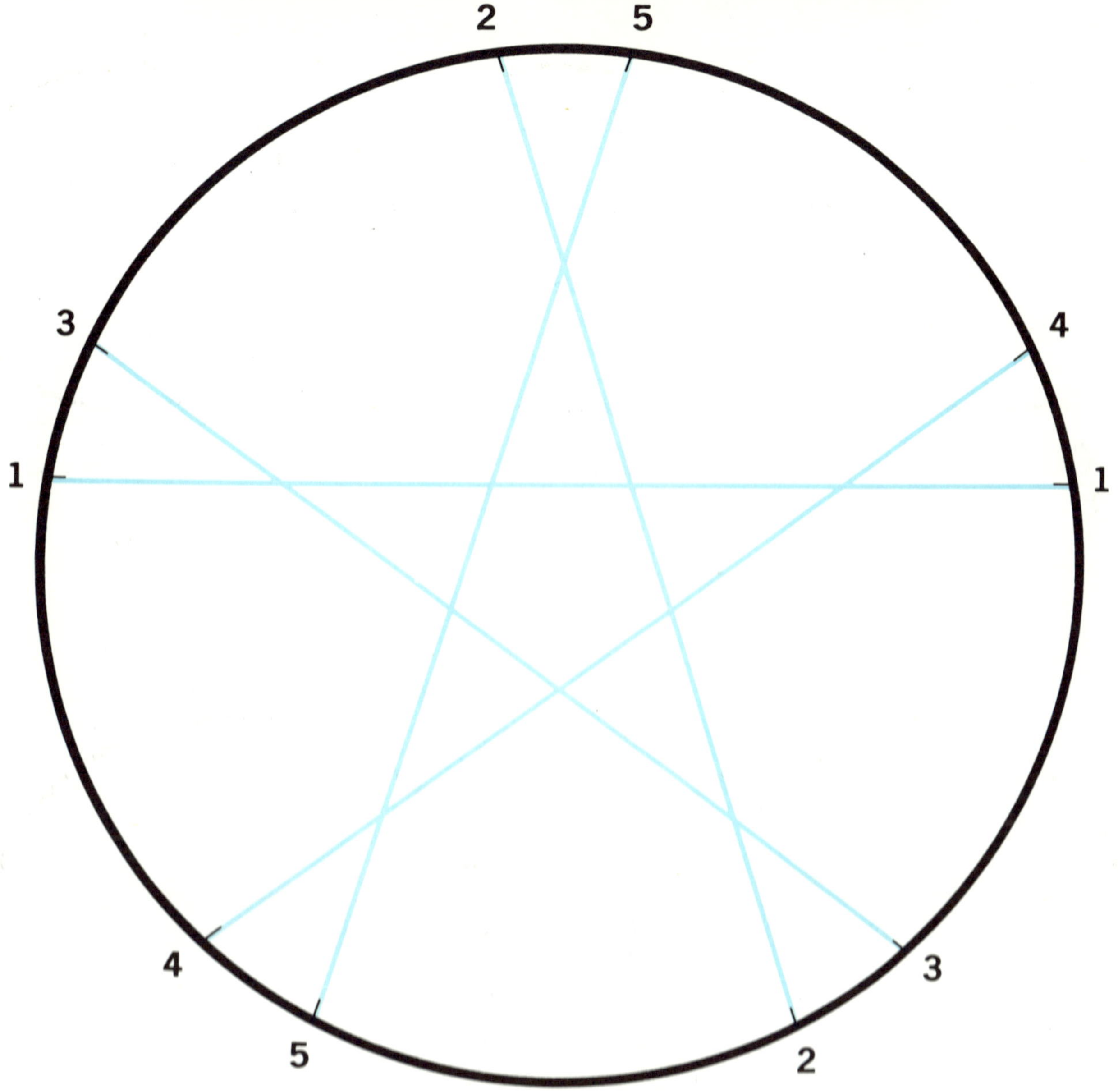

Number of Sections

No. 1	No. 2	No. 3	No. 4	No. 5	No. 6	No. 7	No. 8
2	4	7	11	16	22	29	37

The N th number of sections can be found by using the formula $\frac{N(N+1)}{2} + 1$

Circle Sections **B**

CIRCLE SECTIONS

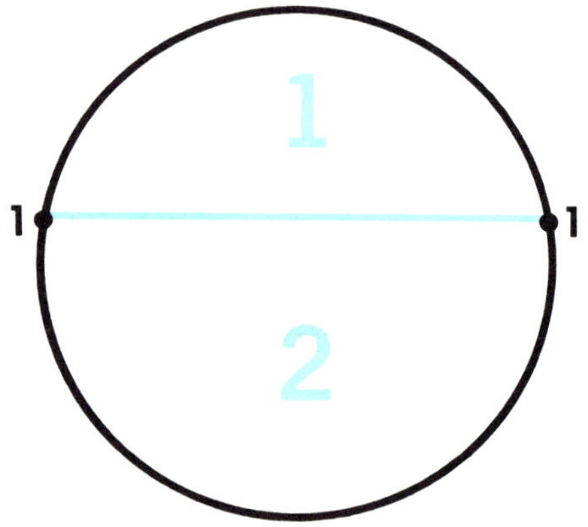

Cut No. 1

1 + 1

2

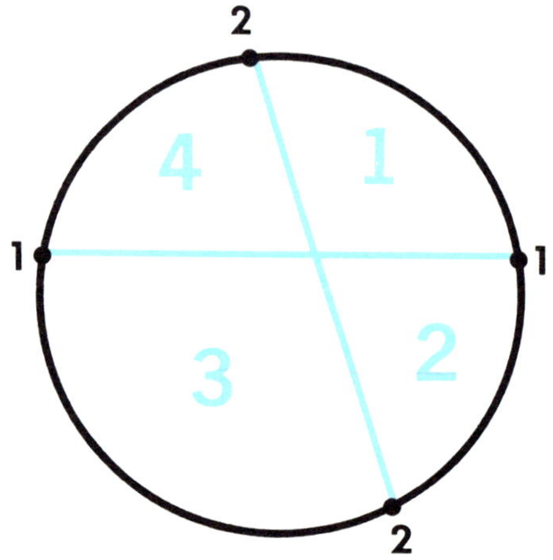

Cut No. 2

2 + 2

4

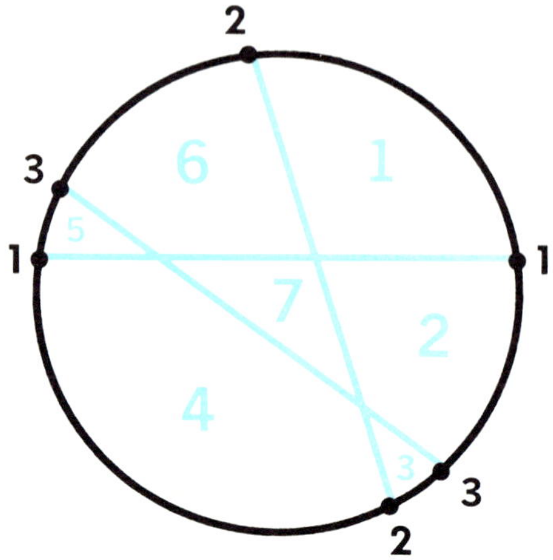

Cut No. 3

4 + 3

7

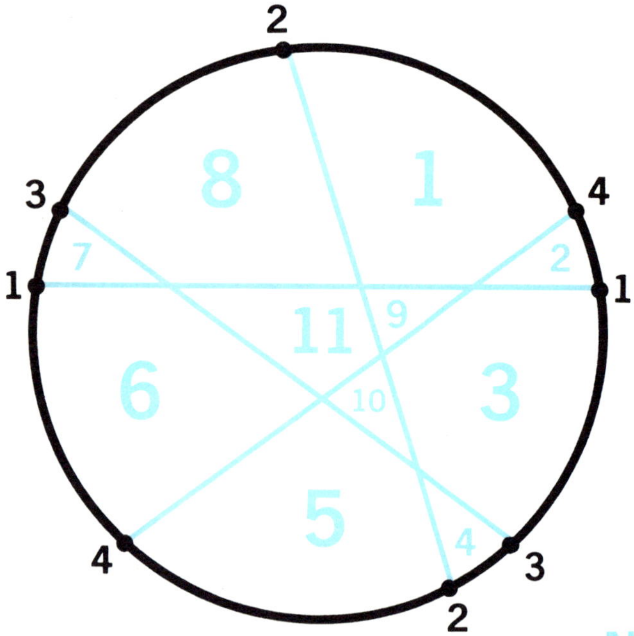

Cut No. 4

7 + 4

11

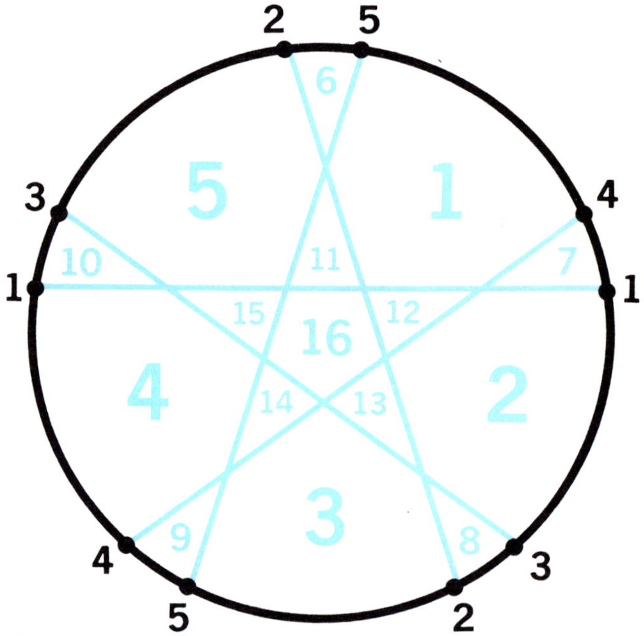

Cut No. 5

11 + 5

16

Notice

Each cut must intersect all previous cuts to find the greatest number of regions to be formed.

Circle Sections **A**

NARCISSISTIC NUMBERS

According to an ancient Greek Myth the beautiful youth Narcissus fell in love with his own reflection as he bent over a clear pool to drink. The goddess of righteous anger changed him into the flower which bears his name, thus leaving him leaning forever gazing into the pool. Hence from this myth comes the name Narcissistic given to numbers which can be represented as some function of its digits.

Form

A) $2427 = 2^1 + 4^2 + 2^3 + 7^4$

B) $1676 = 1^5 + 6^4 + 7^3 + 6^2$

C) $40585 = 4! + 0! + 5! + 8! + 5!$

Prove which of the following numbers are Narcissistic.

Form A	Form B	Form C
A) 135	E) 24	I) 135
B) 598	F) 89	J) 145
C) 6765	G) 332	K) 3435
D) 47016	H) 1676	L) 40585

(C,F,I & K are not Narcissistic)

PRIME NUMBERS

Any natural number greater than one which can be divided evenly by only the number itself and one.

2, 3, 5, 7, 11, 13, 17, 19...

Use the Sieve of Eratosthenes to find primes less than 100.

TESTING LARGE NUMBERS FOR PRIMALITY

Example: Is 1,247 a prime number?

Use the table of Squares and Square Roots to find the largest square number less than 1,247.

$1,296 = (36)^2$

1,247 — Prime numbers less than 35

$1,225 = (35)^2$ — 31, 29, 23, 19, 17, 13, 11, 7, 5, 3, 2.

If the above prime numbers are not proper divisors of 1,247, then the number is prime.

COMPOSITE NUMBERS

A natural number that can be expressed as a product of two or more prime factors in one and only one way.

4, 6, 8, 9, 10, 12, 14, 15...

The prime factors of a composite number should be arranged in ascending order from left to right.

$$12 = 2 \times 2 \times 3 \text{ or } (2)^2\, 3$$

SIEVE OF ERATOSTHENES

A similar chart was used by the mathematician Eratosthenes (276–196 B.C.) to find the prime numbers from 1 to 100.

1	2	3	4	5	6	7	8	9	10
11	12	13	14	15	16	17	18	19	20
21	22	23	24	25	26	27	28	29	30
31	32	33	34	35	36	37	38	39	40
41	42	43	44	45	46	47	48	49	50
51	52	53	54	55	56	57	58	59	60
61	62	63	64	65	66	67	68	69	70
71	72	73	74	75	76	77	78	79	80
81	82	83	84	85	86	87	88	89	90
91	92	93	94	95	96	97	98	99	100

Sieve of Eratosthenes

SIEVE OF ERATOSTHENES

PRIME NUMBERS

1	2	3	4	5	6	7	8	9	10
11	12	13	14	15	16	17	18	19	20
21	22	23	24	25	26	27	28	29	30
31	32	33	34	35	36	37	38	39	40
41	42	43	44	45	46	47	48	49	50
51	52	53	54	55	56	57	58	59	60
61	62	63	64	65	66	67	68	69	70
71	72	73	74	75	76	77	78	79	80
81	82	83	84	85	86	87	88	89	90
91	92	93	94	95	96	97	98	99	100

(25 Primes)

Transparency Overlay

PRIME NUMBER SPIRAL

Beginning in the center with 11, continue numbering throughout the spiral until you have reached 110.

110	109	108	107	106	105	104	103	102	101
75	74	73	72	71	70	69	68	67	100
76	47	46	45	44	43	42	41	66	99
77	48	27	26	25	24	23	40	65	98
78	49	28	15	14	13	22	39	64	97
79	50	29	16	11	12	21	38	63	96
80	51	30	17	18	19	20	37	62	95
81	52	31	32	33	34	35	36	61	94
82	53	54	55	56	57	58	59	60	93
83	84	85	86	87	88	89	90	91	92

Circle all prime numbers starting with 11. Did you discover a pattern of prime numbers? If you continue the spiral, will the pattern be repeated indefinitely?

PRIME NUMBER SPIRAL

110	109	108	107	106	105	104	103	102	101
75	74	73	72	71	70	69	68	67	100
76	47	46	45	44	43	42	41	66	99
77	48	27	26	25	24	23	40	65	98
78	49	28	15	14	13	22	39	64	97
79	50	29	16	11	12	21	38	63	96
80	51	30	17	18	19	20	37	62	95
81	52	31	32	33	34	35	36	61	94
82	53	54	55	56	57	58	59	60	93
83	84	85	86	87	88	89	90	91	92

To be used as a separate activity prior to this lesson to encourage students to discover the number pattern.

Transparency Overlay

PRIME NUMBER SPIRAL

Start circling primes in the center of the spiral, rather than at random, and continue working through the spiral.

110	109	108	107	106	105	104	103	102	101
75	74	73	72	71	70	69	68	67	100
76	47	46	45	44	43	42	41	66	99
77	48	27	26	25	24	23	40	65	98
78	49	28	15	14	13	22	39	64	97
79	50	29	16	11	12	21	38	63	96
80	51	30	17	18	19	20	37	62	95
81	52	31	32	33	34	35	36	61	94
82	53	54	55	56	57	58	59	60	93
83	84	85	86	87	88	89	90	91	92

Is the same diagonal pattern produced if different center prime numbers are used? Such as prime numbers 7, 13, 17, etc.

If 41 is used, a diagonal of 40 primes are generated in a 40×40 square.

Transparency Overlay

PRIME NUMBER SPIRAL

It is recommended that the teacher indicate to the students the first and the last number of the spiral.

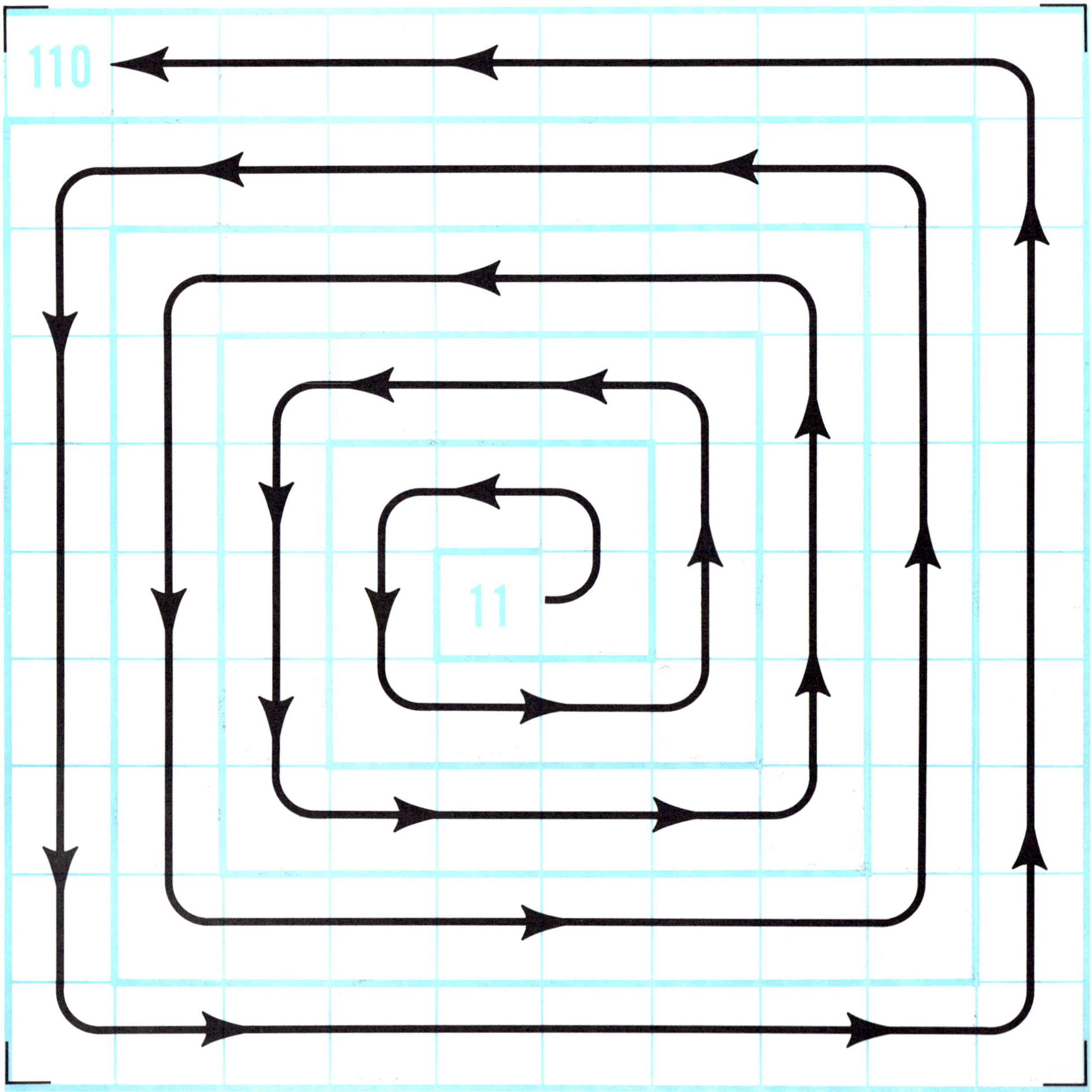

Some students may mistakenly omit a number in the spiral, causing a break in the pattern.

Transparency Overlay

PRIMES AND COMPOSITES

Factorizations

1 = Special	2 = Prime	3 = Prime	4 = 2 x 2
5 = Prime	6 = 2 x 3	7 = Prime	8 = 2 x 2 x 2
9 = 3 x 3	10 = 2 x 5	11 = Prime	12 = 2 x 2 x 3
13 = Prime	14 = 2 x 7	15 = 3 x 5	16 = $(2)^4$
17 = Prime	18 = 2 x 3 x 3	19 = Prime	20 = 2 x 2 x 5
21 = 3 x 7	22 = 2 x 11	23 = Prime	24 = $(2)^3 3$
25 = 5 x 5	26 = 2 x 13	27 = 3 x 3 x 3	28 = 2 x 2 x 7
29 = Prime	30 = 2 x 3 x 5	31 = Prime	32 = $(2)^5$
33 = 3 x 11	34 = 2 x 17	35 = 5 x 7	36 = $(2)^2 (3)^2$
37 = Prime	38 = 2 x 19	39 = 3 x 13	40 = $(2)^3 5$
41 = Prime	42 = 2 x 3 x 7	43 = Prime	44 = 2 x 2 x 11
45 = 3 x 3 x 5	46 = 2 x 23	47 = Prime	48 = $(2)^4 3$
49 = 7 x 7	50 = 2 x 5 x 5	51 = 3 x 17	52 = 2 x 2 x 13
53 = Prime	54 = $2(3)^3$	55 = 5 x 11	56 = $(2)^3 7$
57 = 3 x 19	58 = 2 x 29	59 = Prime	60 = $(2)^2 (3)(5)$

PRIMES AND COMPOSITES

Factorizations

1 = Special	2 = Prime	3 = Prime	4 = 2 x 2
5 = Prime	6 = 2 x 3	7 = Prime	8 = 2 x 2 x 2
9 = 3 x 3	10 = 2 x 5	11 = Prime	12 = 2 x 2 x 3
13 = Prime	14 = 2 x 7	15 = 3 x 5	16 = $(2)^4$
17 = Prime	18 = 2 x 3 x 3	19 = Prime	20 = 2 x 2 x 5
21 = 3 x 7	22 = 2 x 11	23 = Prime	24 = $(2)^3 3$
25 = 5 x 5	26 = 2 x 13	27 = 3 x 3 x 3	28 = 2 x 2 x 7
29 = Prime	30 = 2 x 3 x 5	31 = Prime	32 = $(2)^5$
33 = 3 x 11	34 = 2 x 17	35 = 5 x 7	36 = $(2)^2 (3)^2$
37 = Prime	38 = 2 x 19	39 = 3 x 13	40 = $(2)^3 5$
41 = Prime	42 = 2 x 3 x 7	43 = Prime	44 = 2 x 2 x 11
45 = 3 x 3 x 5	46 = 2 x 23	47 = Prime	48 = $(2)^4 3$
49 = 7 x 7	50 = 2 x 5 x 5	51 = 3 x 17	52 = 2 x 2 x 13
53 = Prime	54 = $2(3)^3$	55 = 5 x 11	56 = $(2)^3 7$
57 = 3 x 19	58 = 2 x 29	59 = Prime	60 = $(2)^2(3)(5)$

Transparency Overlay

61 = Prime	62 = 2 x 31	63 = $(3)^2\,7$	64 = $(2)^6$
65 = 5 x 13	66 = 2 x 3 x 11	67 = Prime	68 = $(2)^2\,17$
69 = 3 x 23	70 = 2 x 5 x 7	71 = Prime	72 = $(2)^3\,(3)^2$
73 = Prime	74 = 2 x 37	75 = $3\,(5)^2$	76 = $(2)^2\,19$
77 = 7 x 11	78 = 2 x 3 x 13	79 = Prime	80 = $(2)^4\,5$
81 = $(3)^4$	82 = 2 x 41	83 = Prime	84 = $(2)^2(3)(7)$
85 = 5 x 17	86 = 2 x 43	87 = 3 x 29	88 = $(2)^3\,11$
89 = Prime	90 = $2\,(3)^2\,5$	91 = 7 x 13	92 = $(2)^2\,23$
93 = 3 x 31	94 = 2 x 47	95 = 5 x 19	96 = $(2)^5\,3$
97 = Prime	98 = $2\,(7)^2$	99 = $(3)^2\,11$	100 = $(2)^2(5)^2$
101 = Prime	102 = 2 x 3 x 17	103 = Prime	104 = $(2)^3\,13$
105 = 3 x 5 x 7	106 = 2 x 53	107 = Prime	108 = $(2)^2\,(3)^3$
109 = Prime	110 = 2 x 5 x 11	111 = 3 x 37	112 = $(2)^4\,7$
113 = Prime	114 = 2 x 3 x 19	115 = 5 x 23	116 = $(2)^2\,29$
117 = $(3)^2\,13$	118 = 2 x 59	119 = 7 x 17	120 = $(2)^3(3)(5)$
121 = $(11)^2$	122 = 2 x 61	123 = 3 x 41	124 = $(2)^2\,31$
125 = $(5)^3$	126 = $2\,(3)^2\,7$	127 = Prime	128 = $(2)^7$
129 = 3 x 43	130 = 2 x 5 x 13	131 = Prime	132 = $(2)^2(3)(11)$

61 = Prime	62 = 2 x 31	63 = $(3)^2\,7$	64 = $(2)^6$
65 = 5 x 13	66 = 2x3x11	67 = Prime	68 = $(2)^2\,17$
69 = 3 x 23	70 = 2x5x7	71 = Prime	72 = $(2)^3\,(3)^2$
73 = Prime	74 = 2 x 37	75 = $3\,(5)^2$	76 = $(2)^2\,19$
77 = 7 x 11	78 = 2x3x13	79 = Prime	80 = $(2)^4\,5$
81 = $(3)^4$	82 = 2 x 41	83 = Prime	84 = $(2)^2(3)(7)$
85 = 5 x 17	86 = 2 x 43	87 = 3 x 29	88 = $(2)^3\,11$
89 = Prime	90 = $2\,(3)^2\,5$	91 = 7 x 13	92 = $(2)^2\,23$
93 = 3 x 31	94 = 2 x 47	95 = 5 x 19	96 = $(2)^5\,3$
97 = Prime	98 = $2\,(7)^2$	99 = $(3)^2\,11$	100 = $(2)^2(5)^2$
101 = Prime	102 = 2x3x17	103 = Prime	104 = $(2)^3\,13$
105 = 3x5x7	106 = 2 x 53	107 = Prime	108 = $(2)^2\,(3)^3$
109 = Prime	110 = 2x5x11	111 = 3 x 37	112 = $(2)^4\,7$
113 = Prime	114 = 2x3x19	115 = 5 x 23	116 = $(2)^2\,29$
117 = $(3)^2\,13$	118 = 2 x 59	119 = 7 x 17	120 = $(2)^3(3)(5)$
121 = $(11)^2$	122 = 2 x 61	123 = 3 x 41	124 = $(2)^2\,31$
125 = $(5)^3$	126 = $2\,(3)^2\,7$	127 = Prime	128 = $(2)^7$
129 = 3 x 43	130 = 2x5x13	131 = Prime	132 = $(2)^2(3)(11)$

Transparency Overlay

GOLDBACH'S CONJECTURE

Every even number greater than 4 is the sum of two odd prime numbers. Example: 8 = 5 + 3

Possible Solutions

6 = 3 + 3	8 = 5 + 3	10 = 5 + 5	12 = 7 + 5
14 = 11 + 3	16 = 13 + 3	18 = 11 + 7	20 = 13 + 7
22 = 17 + 5	24 = 17 + 7	26 = 23 + 3	28 = 17 + 11
30 = 17 + 13	32 = 19 + 13	34 = 17 + 17	36 = 19 + 17
38 = 19 + 19	40 = 23 + 17	42 = 23 + 19	44 = 31 + 13
46 = 29 + 17	48 = 29 + 19	50 = 37 + 13	52 = 29 + 23
54 = 43 + 11	56 = 43 + 13	58 = 41 + 17	60 = 31 + 29
62 = 43 + 19	64 = 47 + 17	66 = 43 + 23	68 = 37 + 31
70 = 47 + 23	72 = 41 + 31	74 = 61 + 13	76 = 53 + 23
78 = 59 + 19	80 = 61 + 19	82 = 71 + 11	84 = 47 + 37
86 = 73 + 13	88 = 47 + 41	90 = 59 + 31	92 = 79 + 13
94 = 83 + 11	96 = 53 + 43	98 = 61 + 37	100 = 53 + 47
102 = 59 + 43	104 = 67 + 37	106 = 89 + 17	108 = 67 + 41
110 = 73 + 37	112 = 59 + 53	114 = 67 + 47	116 = 79 + 37

More than one solution is possible

GOLDBACH'S CONJECTURE

Every even number greater than 4 is the sum of two odd prime numbers. Example: 8 = 5 + 3

Possible Solutions

6 = 3 + 3	8 = 5 + 3	10 = 5 + 5	12 = 7 + 5
14 = 11 + 3	16 = 13 + 3	18 = 11 + 7	20 = 13 + 7
22 = 17 + 5	24 = 17 + 7	26 = 23 + 3	28 = 17 + 11
30 = 17 + 13	32 = 19 + 13	34 = 17 + 17	36 = 19 + 17
38 = 19 + 19	40 = 23 + 17	42 = 23 + 19	44 = 31 + 13
46 = 29 + 17	48 = 29 + 19	50 = 37 + 13	52 = 29 + 23
54 = 43 + 11	56 = 43 + 13	58 = 41 + 17	60 = 31 + 29
62 = 43 + 19	64 = 47 + 17	66 = 43 + 23	68 = 37 + 31
70 = 47 + 23	72 = 41 + 31	74 = 61 + 13	76 = 53 + 23
78 = 59 + 19	80 = 61 + 19	82 = 71 + 11	84 = 47 + 37
86 = 73 + 13	88 = 47 + 41	90 = 59 + 31	92 = 79 + 13
94 = 83 + 11	96 = 53 + 43	98 = 61 + 37	100 = 53 + 47
102 = 59 + 43	104 = 67 + 37	106 = 89 + 17	108 = 67 + 41
110 = 73 + 37	112 = 59 + 53	114 = 67 + 47	116 = 79 + 37

More than one solution is possible

Transparency Overlay

PRIME NUMBERS
BETWEEN 1 AND 1,000

2	79	191	311	439	577	709	857
3	83	193	313	443	587	719	859
5	89	197	317	449	593	727	863
7	97	199	331	457	599	733	877
11	101	211	337	461	601	739	881
13	103	223	347	463	607	743	883
17	107	227	349	467	613	751	887
19	109	229	353	479	617	757	907
23	113	233	359	487	619	761	911
29	127	239	367	491	631	769	919
31	131	241	373	499	641	773	929
37	137	251	379	503	643	787	937
41	139	257	383	509	647	797	941
43	149	263	389	521	653	809	947
47	151	269	397	523	659	811	953
53	157	271	401	541	661	821	967
59	163	277	409	547	673	823	971
61	167	281	419	557	677	827	977
67	173	283	421	563	683	829	983
71	179	293	431	569	691	839	991
73	181	307	433	571	701	853	997

TWIN PRIMES

Twin primes are consecutive primes that differ by 2.

2	79	191	311	439	577	709	857
3	83	193	313	443	587	719	859
5	89	197	317	449	593	727	863
7	97	199	331	457	599	733	877
11	101	211	337	461	601	739	881
13	103	223	347	463	607	743	883
17	107	227	349	467	613	751	887
19	109	229	353	479	617	757	907
23	113	233	359	487	619	761	911
29	127	239	367	491	631	769	919
31	131	241	373	499	641	773	929
37	137	251	379	503	643	787	937
41	139	257	383	509	647	797	941
43	149	263	389	521	653	809	947
47	151	269	397	523	659	811	953
53	157	271	401	541	661	821	967
59	163	277	409	547	673	823	971
61	167	281	419	557	677	827	977
67	173	283	421	563	683	829	983
71	179	293	431	569	691	839	991
73	181	307	433	571	701	853	997

(35 Twin Primes)

Transparency Overlay

GENERATING NUMBER PATTERNS

1,2	2,3	3,4	4,5	5,6	6,7	7,8
1+2	2+3	3+4	4+5	5+6	6+7	7+8
3	5	7	9	11	13	15

These are odd *numbers.*

1,2,3	2,3,4	3,4,5	4,5,6	5,6,7
1+2+3	2+3+4	3+4+5	4+5+6	5+6+7
6	9	12	15	18

These are multiples of three.

1,2,3,4	2,3,4,5	3,4,5,6	4,5,6,7
Sum is	Sum is	Sum is	Sum is
10	14	18	22

These are even *numbers.*

1,2,3,4,5	2,3,4,5,6	3,4,5,6,7	4,5,6,7,8
Sum is	Sum is	Sum is	Sum is
15	20	25	30

These are multiples of five.

GAUSS' METHOD OF SUMMATION

Problem — Find the sum of terms from 1 — 10.

1 + 2 + 3 + 4 + 5 + 6 + 7 + 8 + 9 + 10 = _____.

Procedure —

1	2	3	4	5	6	7	8	9	10
+10	9	8	7	6	5	4	3	2	1

11 + 11 + 11 + 11 + 11 + 11 + 11 + 11 + 11 + 11 = 110

* Since each term was used twice, divide the sum 110 by 2 = 55 answer.

Short-cut —

First Term		Last Term		Sum		Number of Terms		Product	*			Sum of Terms
1	+	10	=	11	x	10	=	110	÷	2	=	55

Formula —

$$S = \frac{n(n+1)}{2} = \frac{10(10+1)}{2} = 55$$

Exercises —

1. 1 – 15, ans. 120.
2. 1 – 23, ans. 276.
3. 1 – 48, ans. 1,176.
4. 1 – 67, ans. 2,278.
5. 1 – 82, ans. 3,403.
6. 1 – 95, ans. 4,560.

LEAST COMMON MULTIPLE
(L.C.M.)

The smallest whole number in the intersection of two or more sets of multiples.

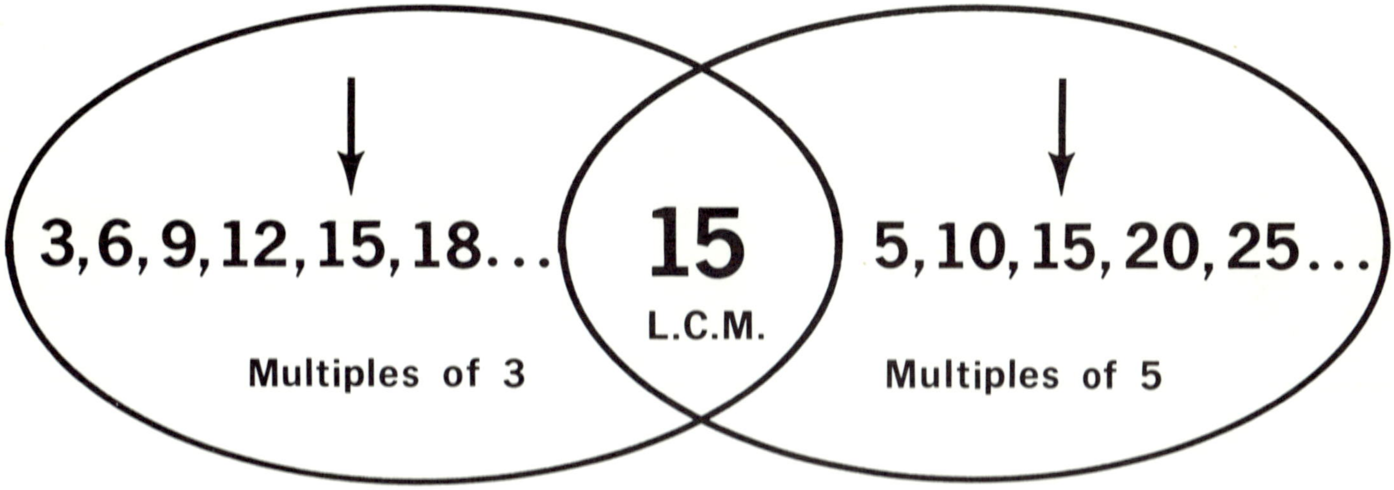

3 = {3, 6, 9, 12, (15,) 18, 21, 24, 27, ...}

L.C.M.

5 = {5, 10, (15,) 20, 25, 30, 35, ...}

The smallest whole number that is divisible by both 3 and 5.

3, 5) L.C.M. = 15

LEAST COMMON MULTIPLE
(L.C.M.)

The smallest whole number that is divisible by each of the two or more members of a given set.

$$\left\{\begin{matrix}18\\28\\40\end{matrix}\right\} \longrightarrow \overline{)\text{L.C.M.}}$$

A. Prime factor each number using exponential notation.

$$\left\{\begin{aligned}18 &= (2)^1\,(3)^2\\ 28 &= (2)^2\,(7)^1\\ 40 &= (2)^3\,(5)^1\end{aligned}\right.$$

B. List one of each different factor used in the above factorization.

C. The exponent of each factor is the largest exponent used on each of the above factors.

$$(2)^3\ (3)^2\ (5)^1\ (7)^1$$

$$8 \cdot 9 \cdot 5 \cdot 7$$

L.C.M. — 2,520

L.C.M. 2

EXPONENTS and POWERS

The exponent indicates how many times the base is used as a factor.

$$\underset{\text{Base}}{(4)}^{3\ \leftarrow\ \text{Exponent}} = \underbrace{4 \times 4 \times 4}_{\text{Factors}} = \underset{\text{Power}}{64}$$

$(4)^0 = 1$
$(4)^1 = 4$ (by agreement)

Note:
The above illustration is usually read as "four to the third power is sixty-four".

$(4)^2 = 4 \times 4 = 16$

$(4)^3 = 4 \times 4 \times 4 = \underline{64}$

$(4)^4 = 4 \times 4 \times \underline{4} \times \underline{4} = \underline{256}$

$(4)^5 = \underline{4} \times \underline{4} \times \underline{4} \times \underline{4} \times \underline{4} = \underline{1,024}$

PASCAL'S TRIANGLE
(Grid)

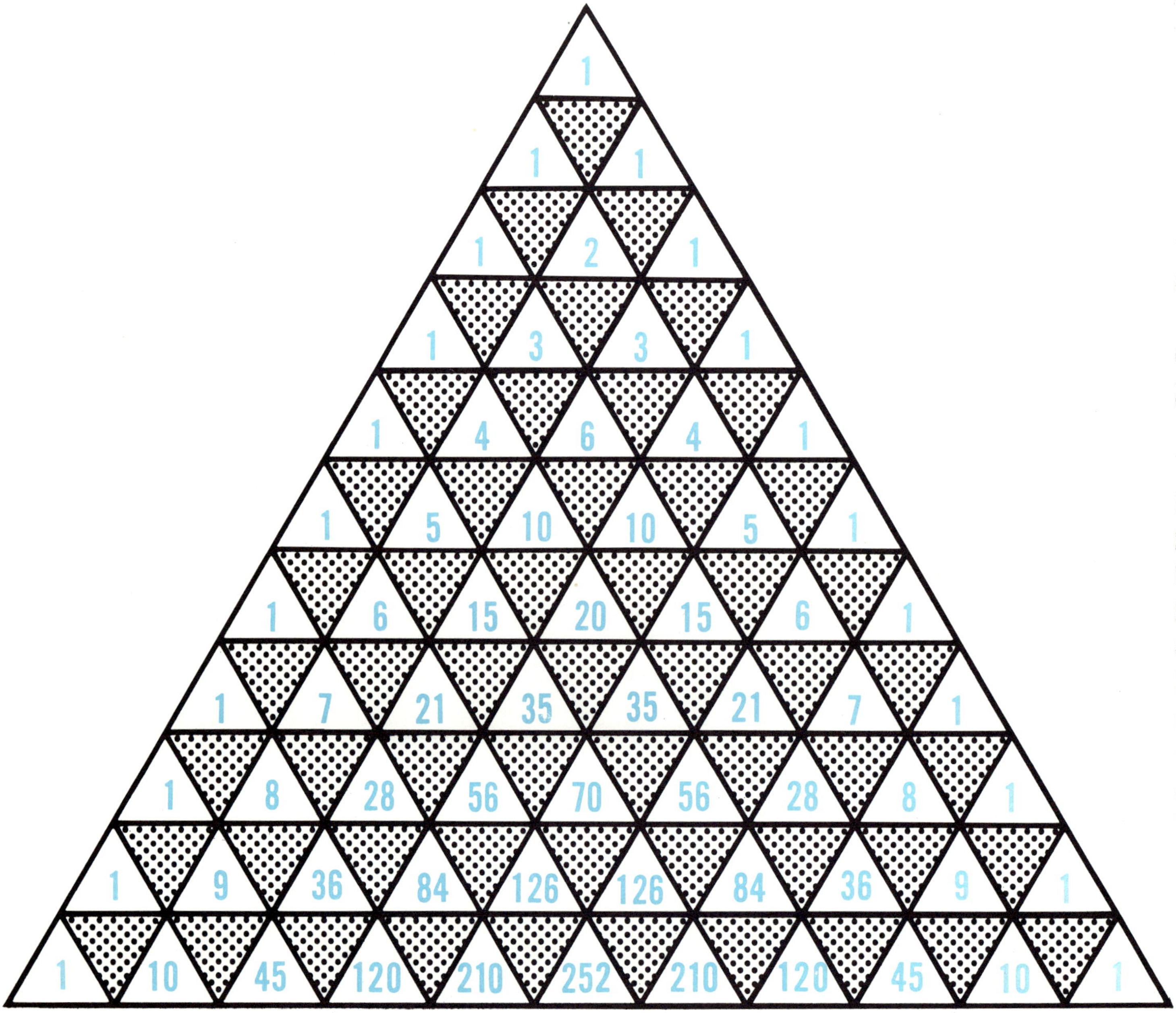

PASCAL'S TRIANGLE

1

1 1

1 2 1

1 3 3 1

1 4 6 4 1

1 5 10 10 5 1

1 6 15 20 15 6 1

1 7 21 35 35 21 7 1

1 8 28 56 70 56 28 8 1

1 9 36 84 126 126 84 36 9 1

1 10 45 120 210 252 210 120 45 10 1

PASCAL'S TRIANGLE

Continuation

1	1	1	1	1	1	1	1
11	12	13	14	15	16	17	18
55	66	78	91	105	120	136	153
165	220	286	364	455	560	680	816
330	495	715	1,001	1,365	1,820	2,380	3,060
462	792	1,287	2,002	3,003	4,368	6,188	8,568
462	924	1,716	3,003	5,005	8,008	12,376	18,564
330	792	1,716	3,432	6,435	11,440	19,448	31,824
165	495	1,287	3,003	6,435	12,870	24,310	43,758
55	220	715	2,002	5,005	11,440	24,310	48,620
11	66	286	1,001	3,003	8,008	19,448	43,758
1	12	78	364	1,365	4,368	12,376	31,824
	1	13	91	455	1,820	6,188	18,564
		1	14	105	560	2,380	8,568
			1	15	120	680	3,060
				1	16	136	816
					1	17	153
						1	18
							1

TRIANGULAR ARRAY

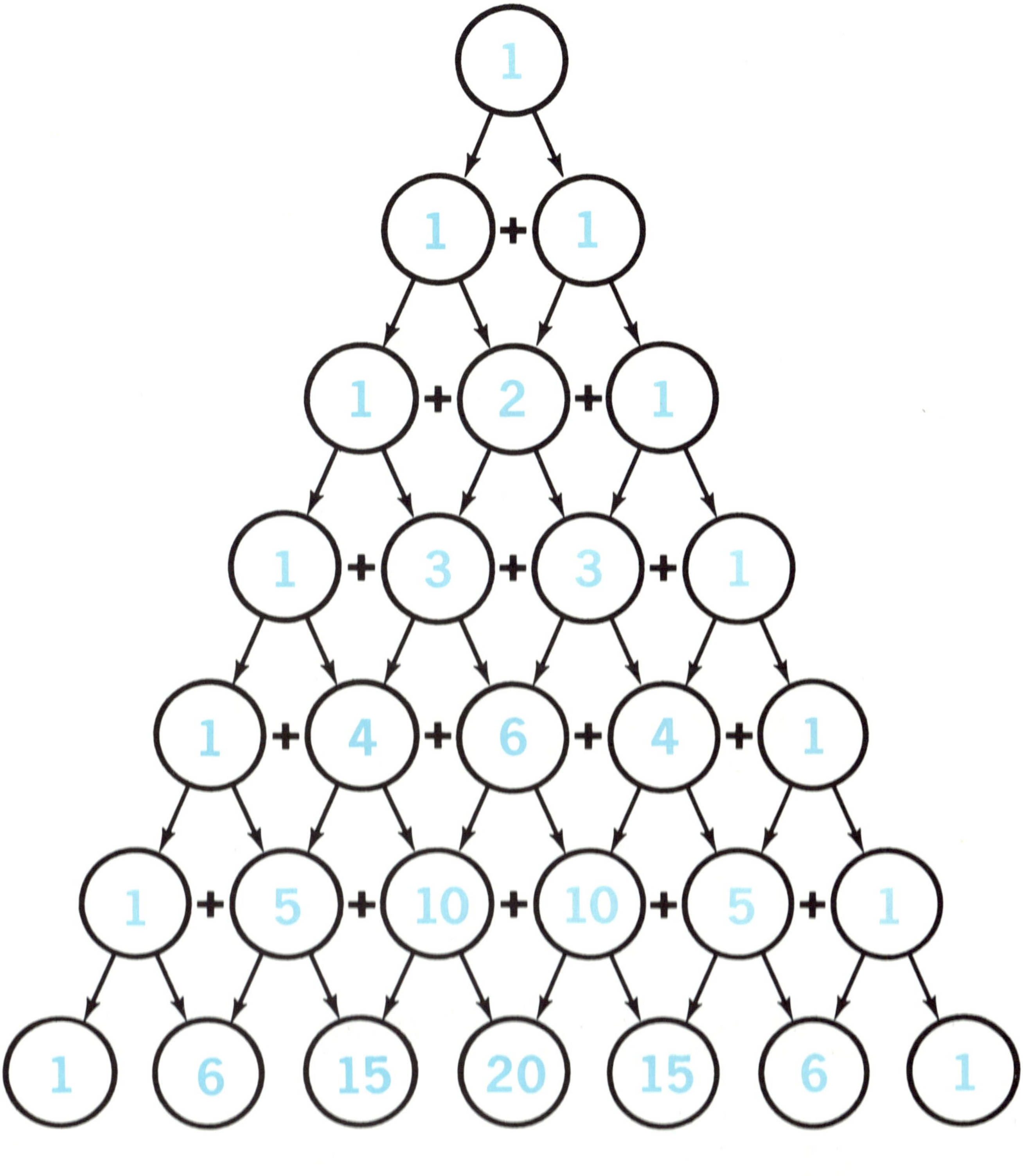

Triangular Array

PASCAL'S TRIANGLE

Row		Sum
$(2)^{0}$	1	1
$(2)^{1}$	1 + 1	2
$(2)^{2}$	1 + 2 + 1	4
$(2)^{3}$	1 + 3 + 3 + 1	8
$(2)^{4}$	1 + 4 + 6 + 4 + 1	16
$(2)^{5}$	1 + 5 + 10 + 10 + 5 + 1	32
$(2)^{6}$	1 + 6 + 15 + 20 + 15 + 6 + 1	64
$(2)^{7}$	1 + 7 + 21 + 35 + 35 + 21 + 7 + 1	128
$(2)^{8}$	1 + 8 + 28 + 56 + 70 + 56 + 28 + 8 + 1	256
$(2)^{9}$	1 + 9 + 36 + 84 + 126 + 126 + 84 + 36 + 9 + 1	512
$(2)^{10}$	1 + 10 + 45 + 120 + 210 + 252 + 210 + 120 + 45 + 10 + 1	1,024

Can you discover the relationship between the number of each row and its related sum?

Pascal's Triangle

PASCAL'S TRIANGLE

TRIANGULAR NUMBERS

Transparency Overlay

PASCAL'S TRIANGLE

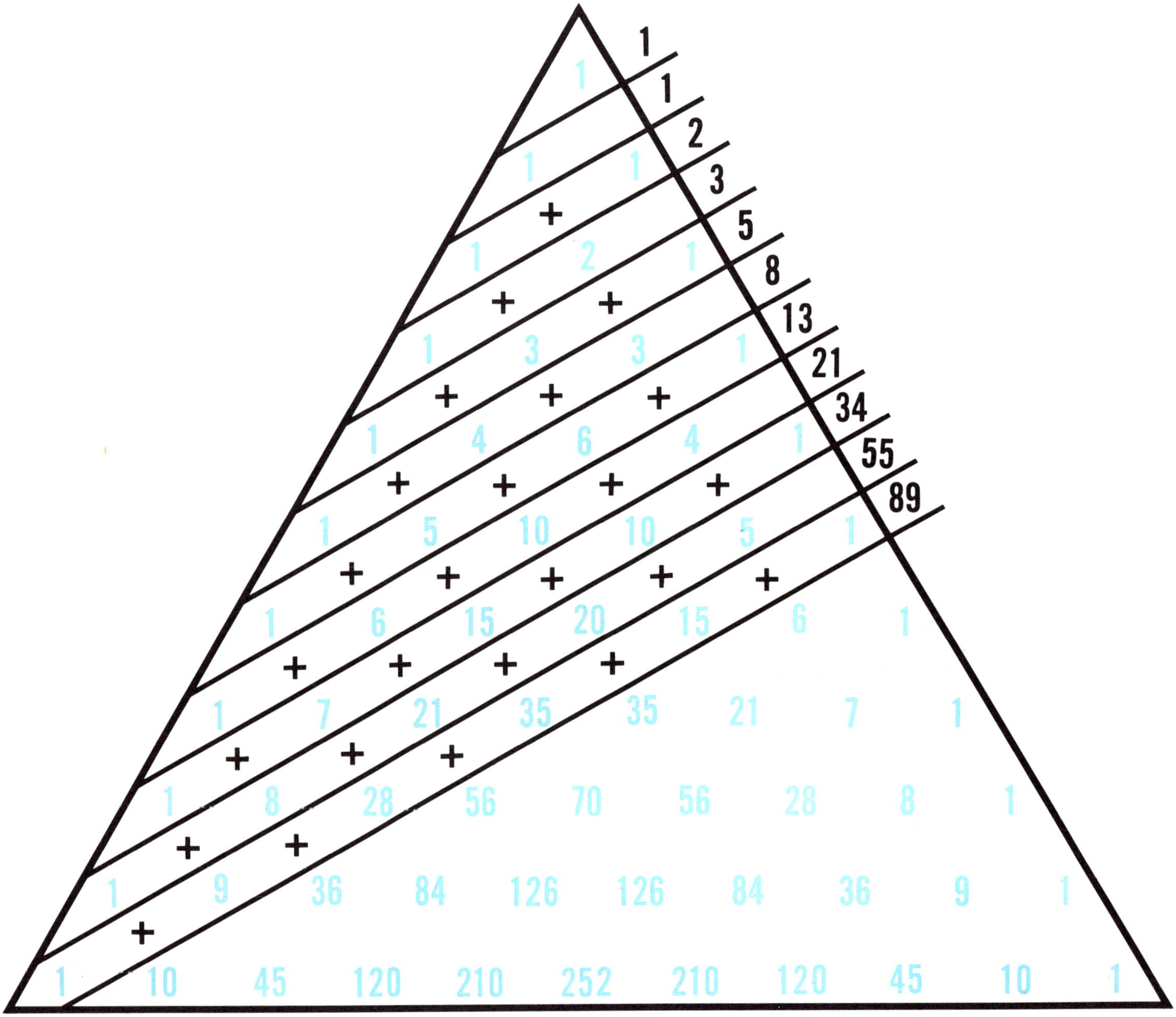

FIBONACCI NUMBERS

Transparency Overlay

Patterns from Pascal's Triangle

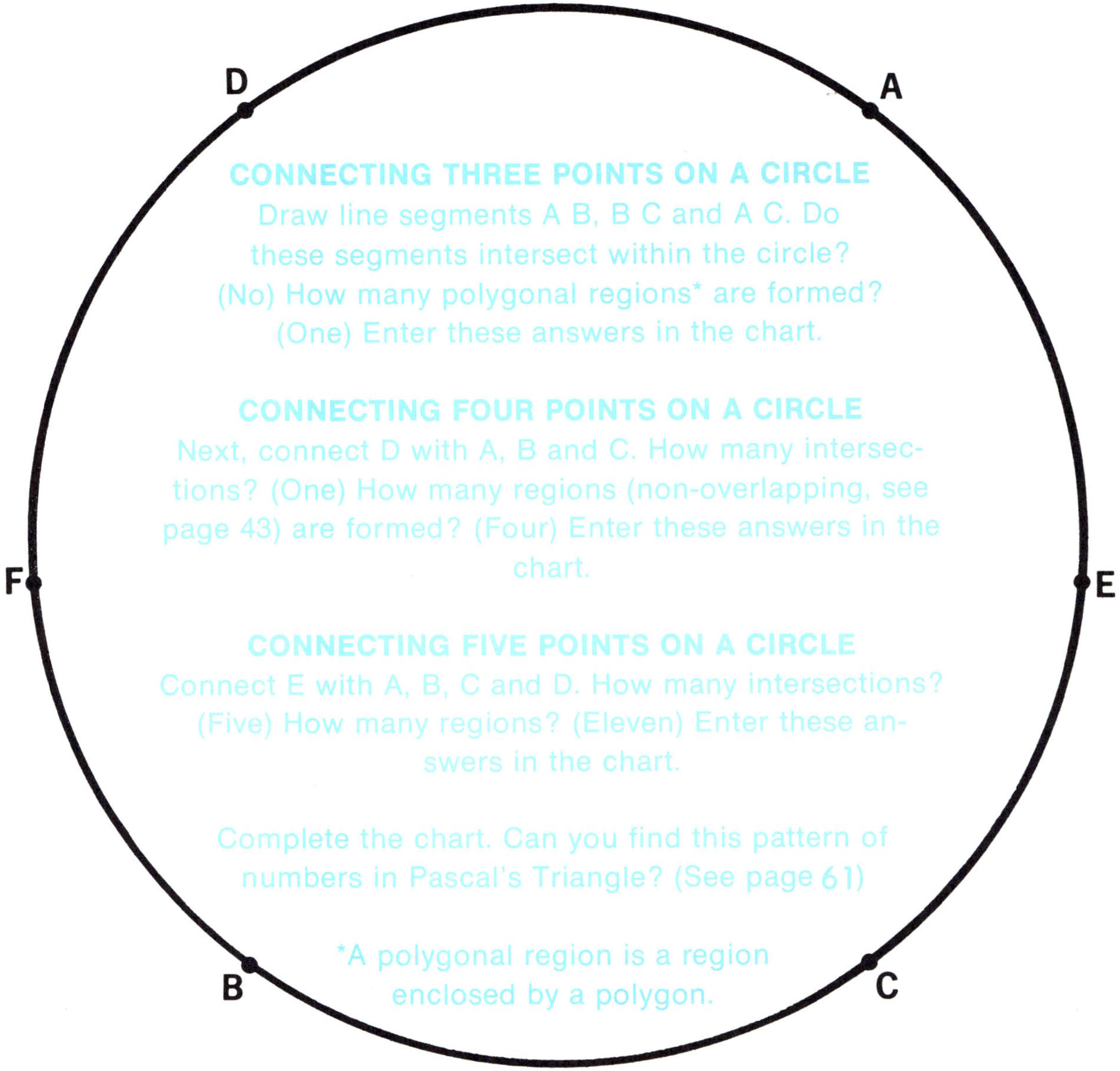

CONNECTING THREE POINTS ON A CIRCLE

Draw line segments A B, B C and A C. Do these segments intersect within the circle? (No) How many polygonal regions* are formed? (One) Enter these answers in the chart.

CONNECTING FOUR POINTS ON A CIRCLE

Next, connect D with A, B and C. How many intersections? (One) How many regions (non-overlapping, see page 43) are formed? (Four) Enter these answers in the chart.

CONNECTING FIVE POINTS ON A CIRCLE

Connect E with A, B, C and D. How many intersections? (Five) How many regions? (Eleven) Enter these answers in the chart.

Complete the chart. Can you find this pattern of numbers in Pascal's Triangle? (See page 61)

*A polygonal region is a region enclosed by a polygon.

Points on a Circle	Number of Regions	Number Intersect Points
3	1	0
4	4	1
5	11	5
6	25	15
7	50	35
8	91	70
9	154	126
10	246	210

Regions and Intersections

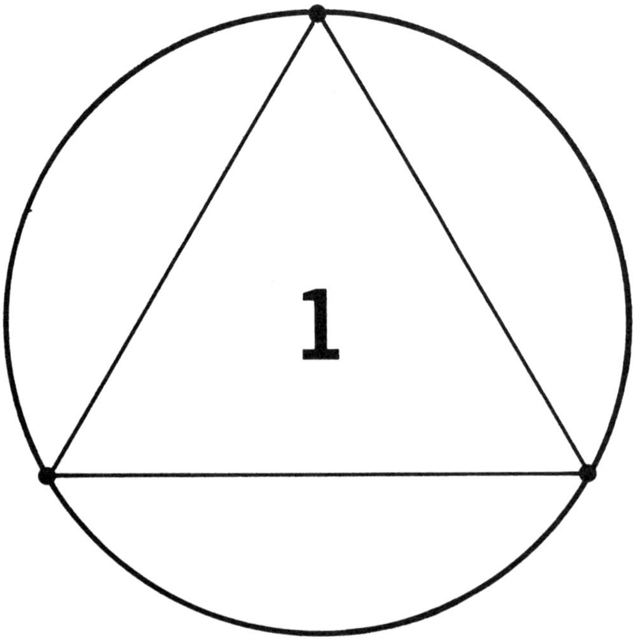

0 Intersections

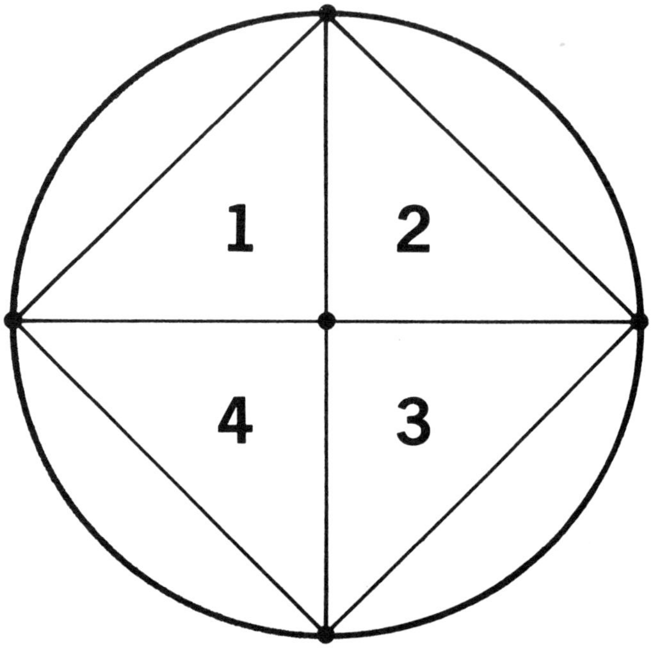

1 Intersection

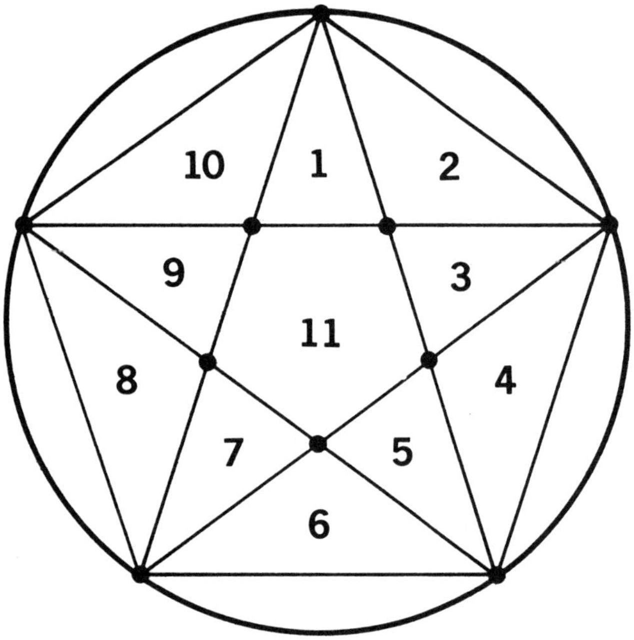

5 Intersections

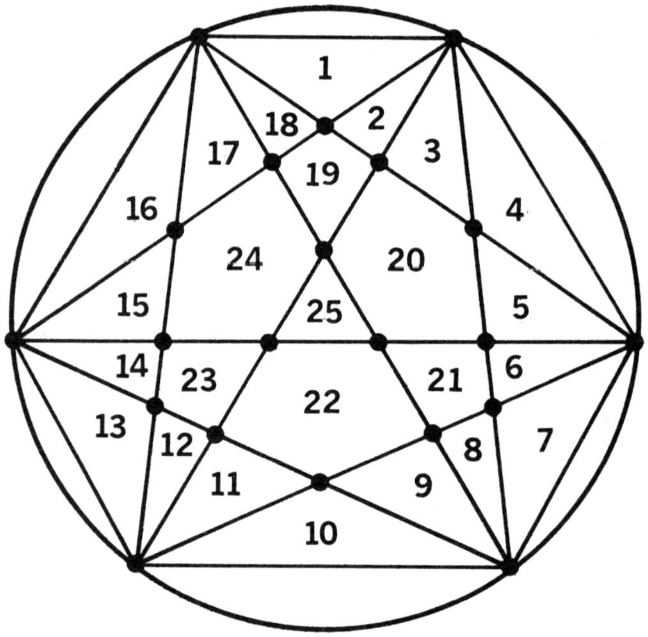

15 Intersections

PASCAL'S TRIANGLE

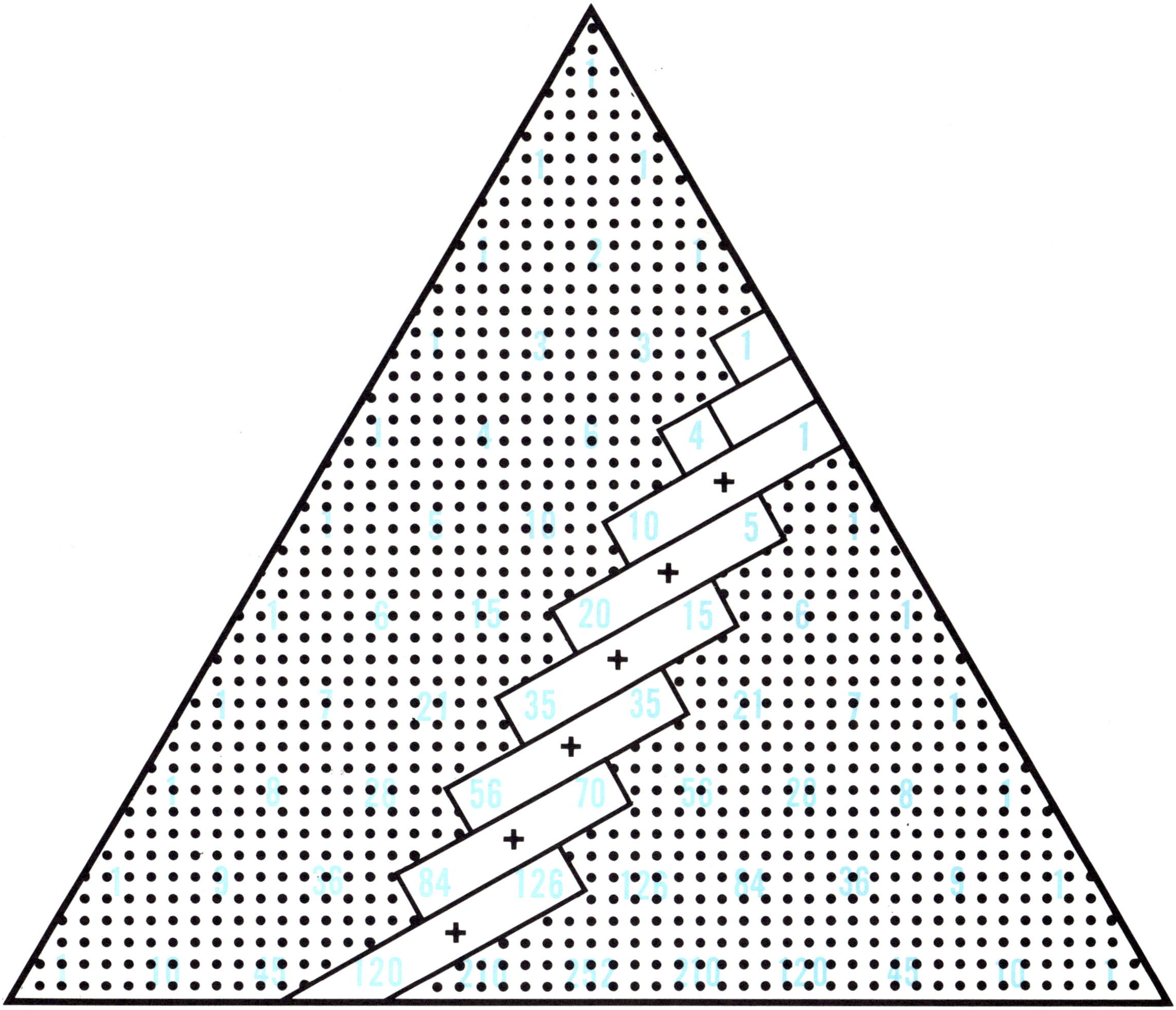

REGIONS IN A CIRCLE

Transparency Overlay

PASCAL'S TRIANGLE

INTERSECTIONS IN A CIRCLE

Transparency Overlay

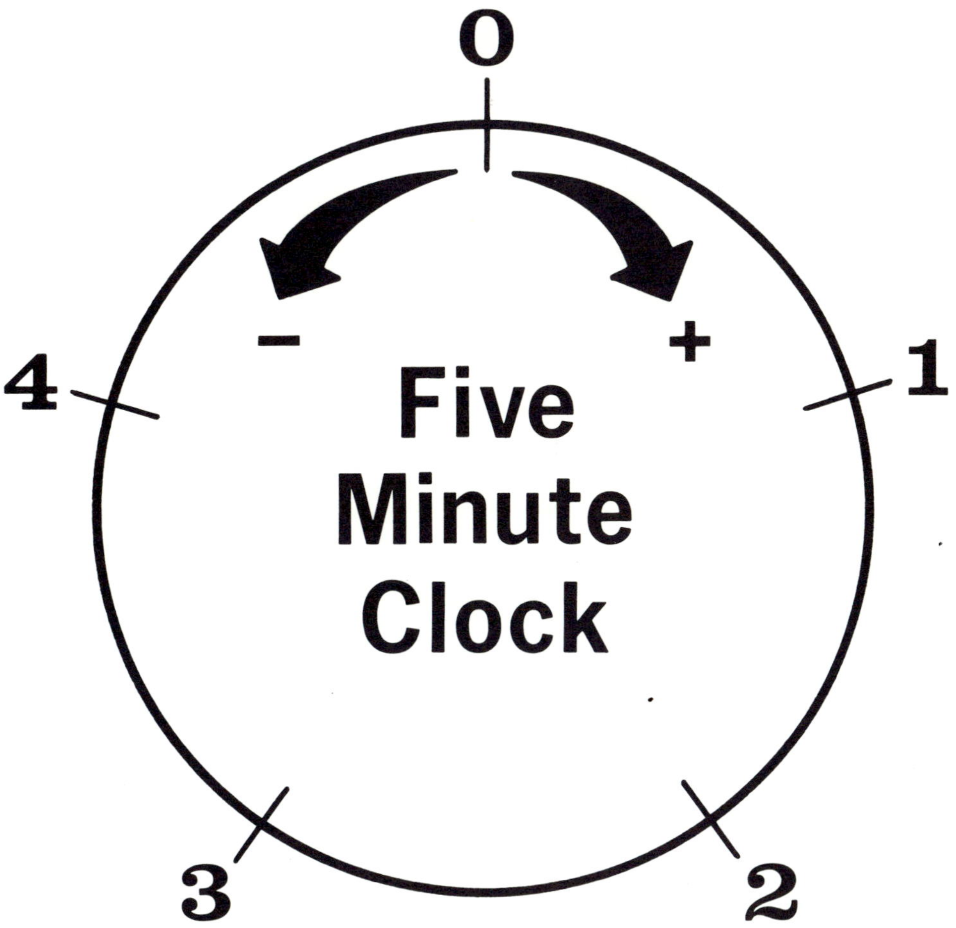

Addition

+	0	1	2	3	4
0	0	1	2	3	4
1	1	2	3	4	0
2	2	3	4	0	1
3	3	4	0	1	2
4	4	0	1	2	3

Multiplication

x	0	1	2	3	4
0	0	0	0	0	0
1	0	1	2	3	4
2	0	2	4	1	3
3	0	3	1	4	2
4	0	4	3	2	1

Addition

+	0	1	2	3	4
0	0	1	2	3	4
1	1	2	3	4	0
2	2	3	4	0	1
3	3	4	0	1	2
4	4	0	1	2	3

Multiplication

x	0	1	2	3	4
0	0	0	0	0	0
1	0	1	2	3	4
2	0	2	4	1	3
3	0	3	1	4	2
4	0	4	3	2	1

Transparency
Overlay

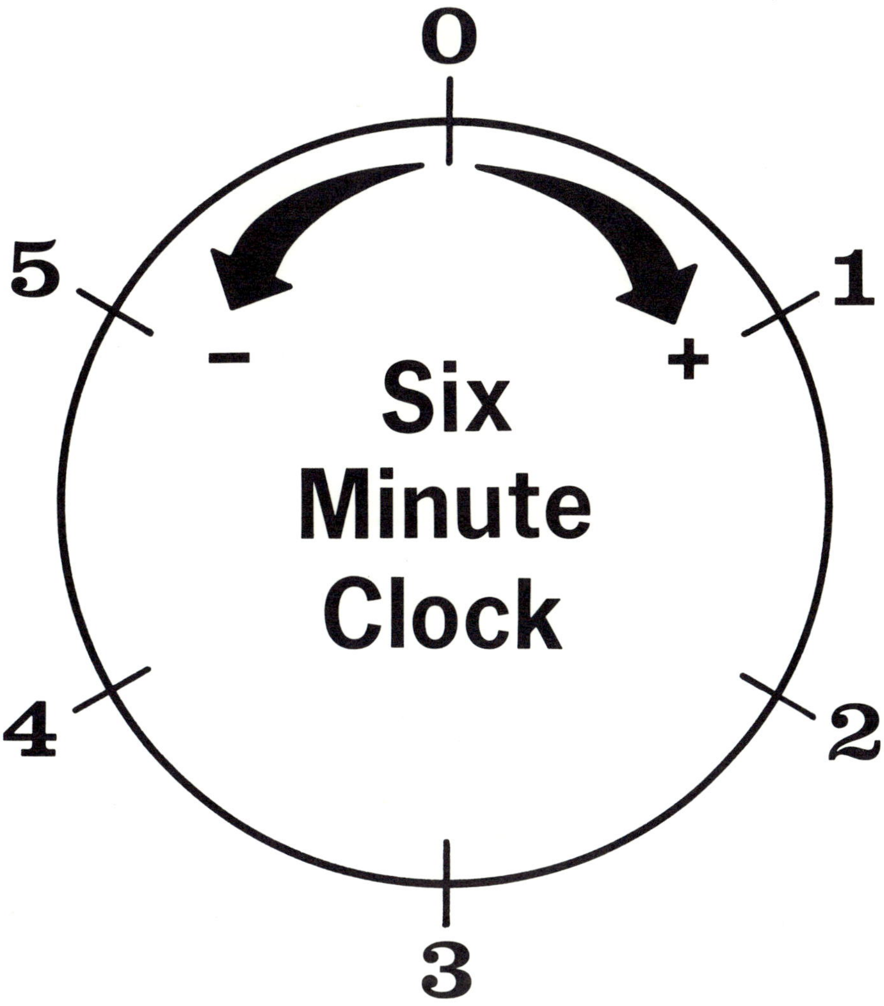

Addition

+	0	1	2	3	4	5
0	0	1	2	3	4	5
1	1	2	3	4	5	0
2	2	3	4	5	0	1
3	3	4	5	0	1	2
4	4	5	0	1	2	3
5	5	0	1	2	3	4

Multiplication

x	0	1	2	3	4	5
0	0	0	0	0	0	0
1	0	1	2	3	4	5
2	0	2	4	0	2	4
3	0	3	0	3	0	3
4	0	4	2	0	4	2
5	0	5	4	3	2	1

Six Minute Clock

Addition

+	0	1	2	3	4	5
0	0	1	2	3	4	5
1	1	2	3	4	5	0
2	2	3	4	5	0	1
3	3	4	5	0	1	2
4	4	5	0	1	2	3
5	5	0	1	2	3	4

Multiplication

x	0	1	2	3	4	5
0	0	0	0	0	0	0
1	0	1	2	3	4	5
2	0	2	4	0	2	4
3	0	3	0	3	0	3
4	0	4	2	0	4	2
5	0	5	4	3	2	1

Transparency Overlay

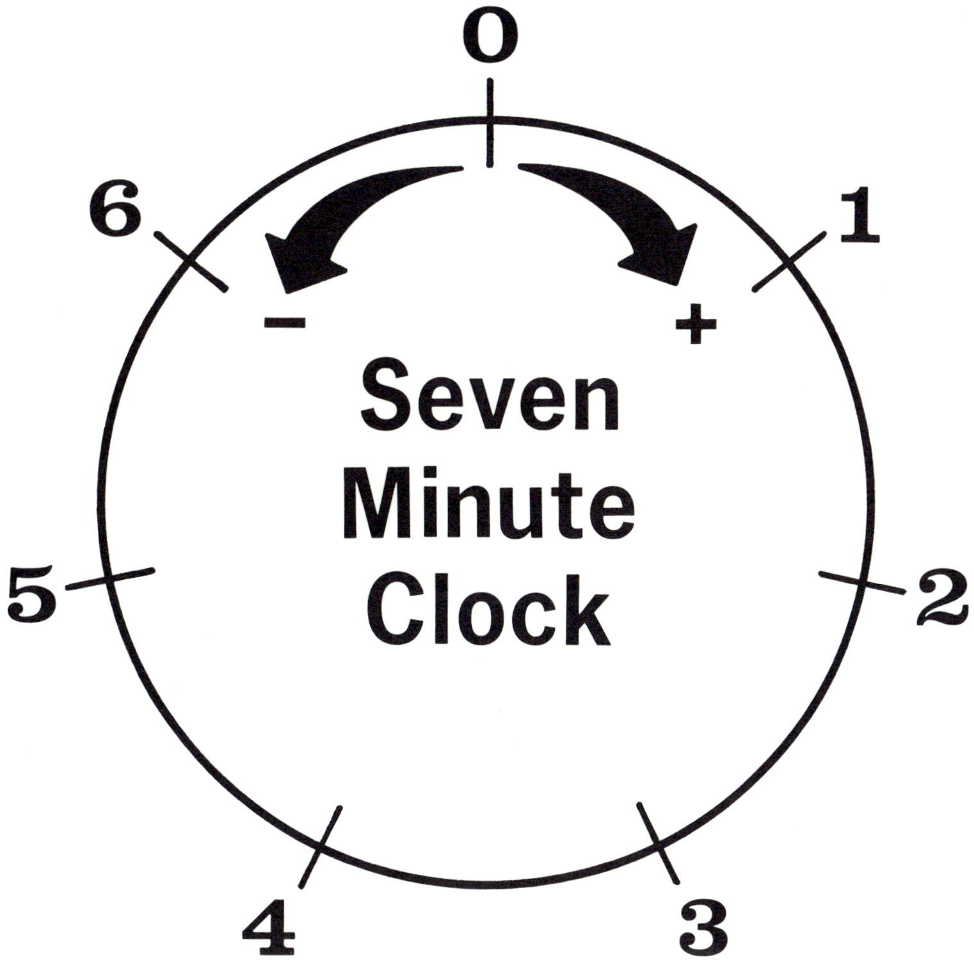

Addition

+	0	1	2	3	4	5	6
0	0	1	2	3	4	5	6
1	1	2	3	4	5	6	0
2	2	3	4	5	6	0	1
3	3	4	5	6	0	1	2
4	4	5	6	0	1	2	3
5	5	6	0	1	2	3	4
6	6	0	1	2	3	4	5

Multiplication

X	0	1	2	3	4	5	6
0	0	0	0	0	0	0	0
1	0	1	2	3	4	5	6
2	0	2	4	6	1	3	5
3	0	3	6	2	5	1	4
4	0	4	1	5	2	6	3
5	0	5	3	1	6	4	2
6	0	6	5	4	3	2	1

Seven Minute Clock

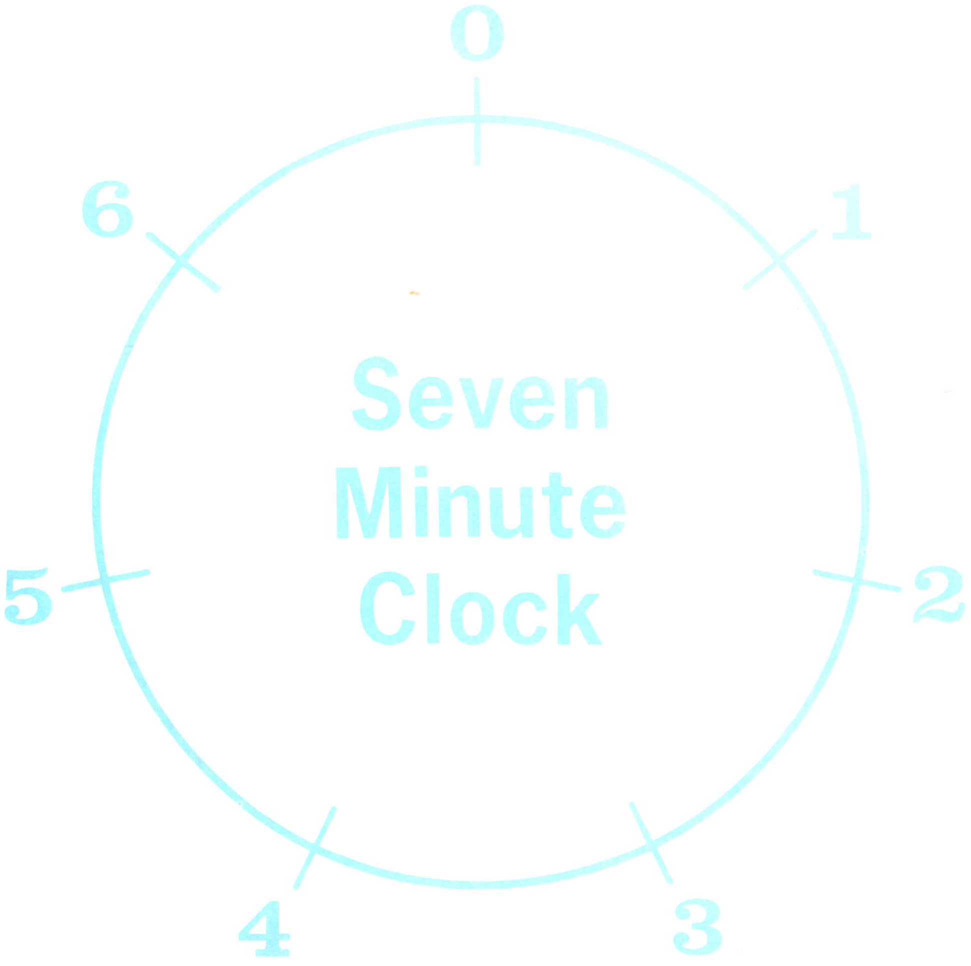

Addition

+	0	1	2	3	4	5	6
0	0	1	2	3	4	5	6
1	1	2	3	4	5	6	0
2	2	3	4	5	6	0	1
3	3	4	5	6	0	1	2
4	4	5	6	0	1	2	3
5	5	6	0	1	2	3	4
6	6	0	1	2	3	4	5

Multiplication

x	0	1	2	3	4	5	6
0	0	0	0	0	0	0	0
1	0	1	2	3	4	5	6
2	0	2	4	6	1	3	5
3	0	3	6	2	5	1	4
4	0	4	1	5	2	6	3
5	0	5	3	1	6	4	2
6	0	6	5	4	3	2	1

Transparency Overlay

THE SQUARE NUMBER THEOREM

(FERMAT)

Every natural number can be expressed as the sum of four squares.

$$327 = (13)^2 + (10)^2 + (7)^2 + (3)^2$$

$$169 + 100 + 49 + 9 = 327$$

$(0)^2$ may be used. Not all solutions are unique.

15	$=(\ 3\ )^2$	+	$(\ 2\ )^2$	+	$(\ 1\ )^2$	+	$(\ 1\ )^2$		
	9	+	4	+	1	+	1	=	15
33	$=(\ 4\ )^2$	+	$(\ 3\ )^2$	+	$(\ 2\ )^2$	+	$(\ 2\ )^2$		
	16	+	9	+	4	+	4	=	33
139	$=(\ 9\ )^2$	+	$(\ 7\ )^2$	+	$(\ 3\ )^2$	+	$(\ 0\ )^2$		
	81	+	49	+	9	+	0	=	139
248	$=(12)^2$	+	$(\ 8\ )^2$	+	$(\ 6\ )^2$	+	$(\ 2\ )^2$		
	144	+	64	+	36	+	4	=	248
868	$=(19)^2$	+	$(17)^2$	+	$(13)^2$	+	$(\ 7\ )^2$		
	361	+	289	+	169	+	49	=	868

Use chart of Squares, Page 83.

RULES OF DIVISIBILITY

A natural number is divisible by:

2 if and only if the units digit of its decimal numeral is 0, 2, 4, 6 or 8, then it is an even number.

3 if and only if the sum of the digits of its decimal numeral is a multiple of 3.

4 if and only if the last two digits of its decimal numeral are divisible by 4.

5 if and only if the units digit of its decimal numeral is 0 or 5.

6 if and only if the decimal numeral is divisible by both 2 and 3.

9 if and only if the sum of the digits of its decimal numeral is a multiple of 9.

10 if and only if the last digit of its decimal numeral is zero.

Divisibility of Natural Numbers

Solutions

		2	3	4	5	6	9	10
1.	24	X	X	X		X		
2.	225		X		X		X	
3.	216	X	X	X		X	X	
4.	7,830	X	X		X	X	X	X
5.	3,753		X				X	
6.	4,566	X	X			X		
7.	20,812	X		X				
8.	537,120	X	X	X	X	X	X	X
9.	413,541		X				X	
10.	185,740	X		X	X			X
11.	16,217		P	R	I	M	E	
12.	135,012	X	X	X		X		
13.	27,285		X		X			
14.	24,105,896	X		X				
15.	318,520	X		X	X			X

(105 Problems)

Divisibility of Natural Numbers

Solutions

		2	3	4	5	6	9	10
1.	24	X	X	X		X		
2.	225		X		X		X	
3.	216	X	X	X		X	X	
4.	7,830	X	X		X	X	X	X
5.	3,753		X				X	
6.	4,566	X	X			X		
7.	20,812	X		X				
8.	537,120	X	X	X	X	X	X	X
9.	413,541		X				X	
10.	185,740	X		X	X			X
11.	16,217		P	R	I	M	E	
12.	135,012	X	X	X		X		
13.	27,285		X		X			
14.	24,105,896	X		X				
15.	318,520	X		X	X			X

(105 Problems)

Transparency Overlay

CASTING OUT NINES

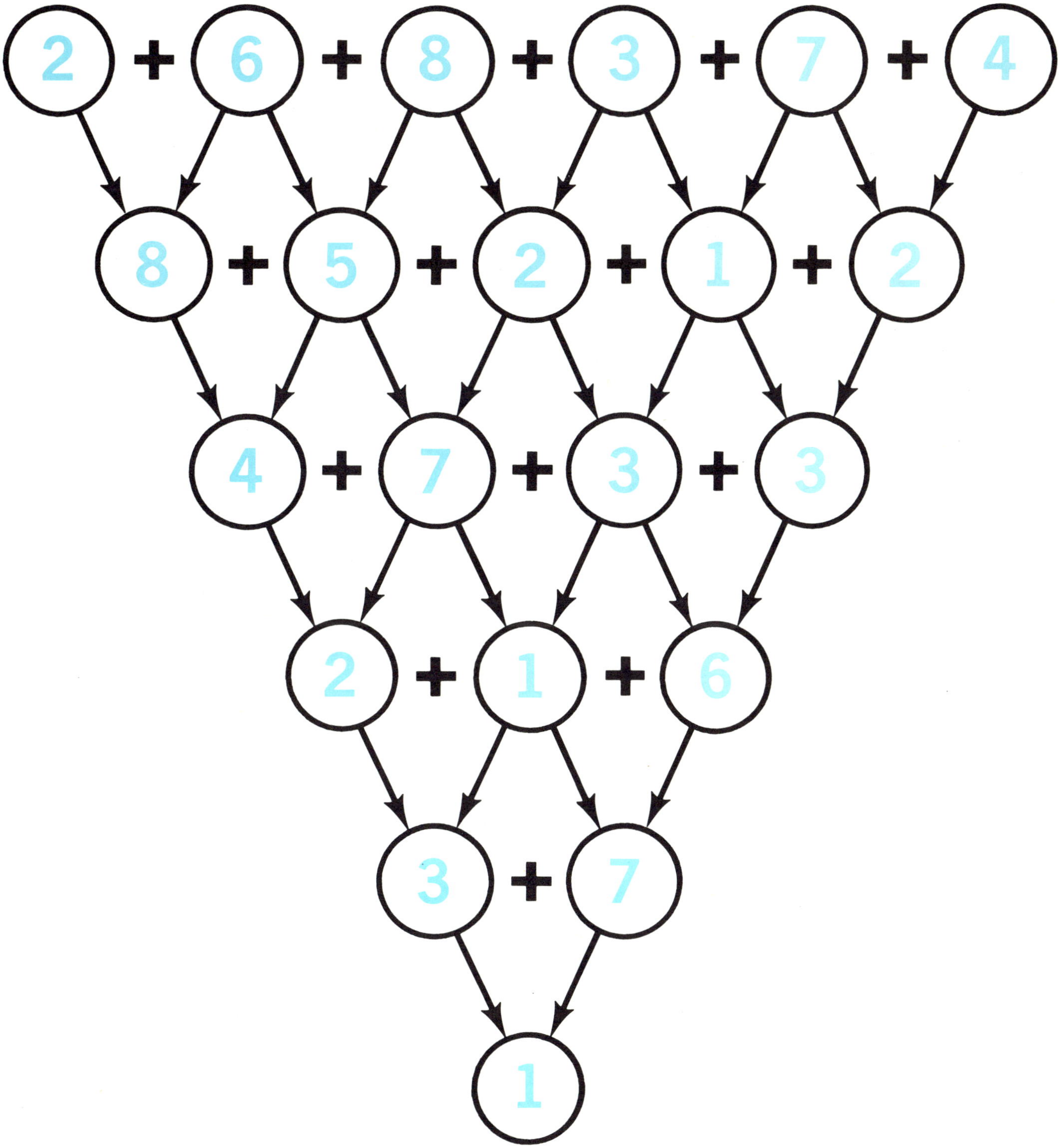

Teacher's Transparency

CASTING OUT NINES

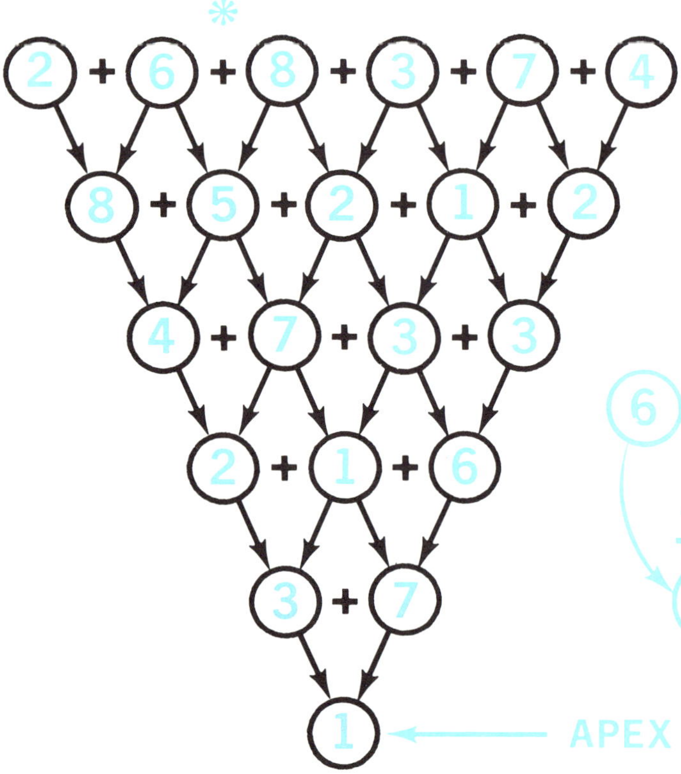

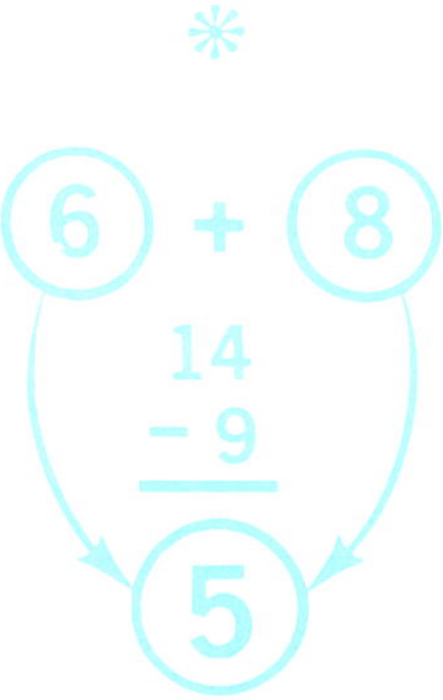

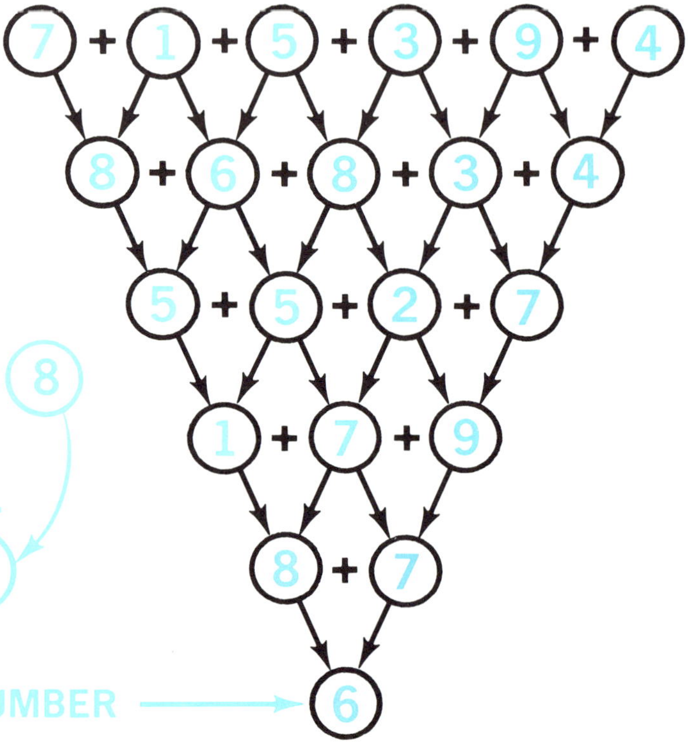

APEX NUMBER

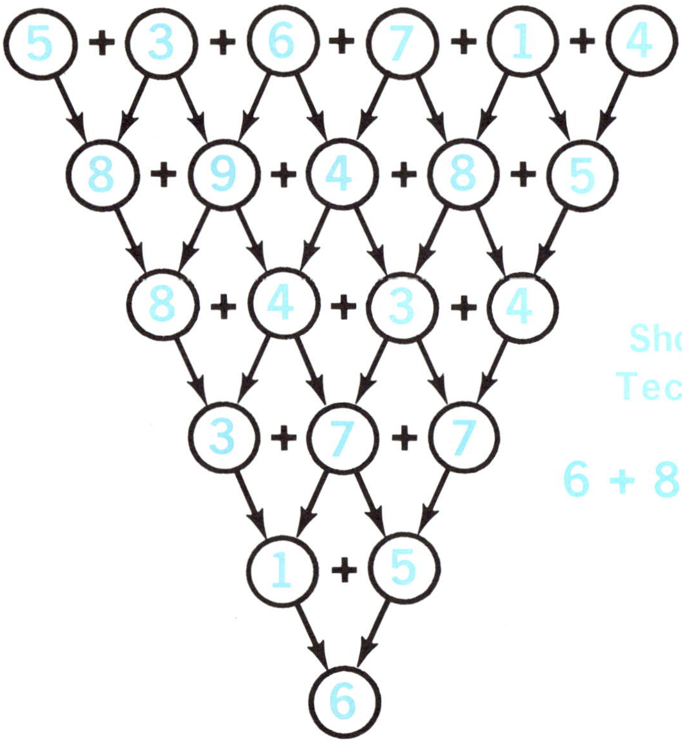

*
Short-cut
Technique

6 + 8 = 14

1+4 = 5

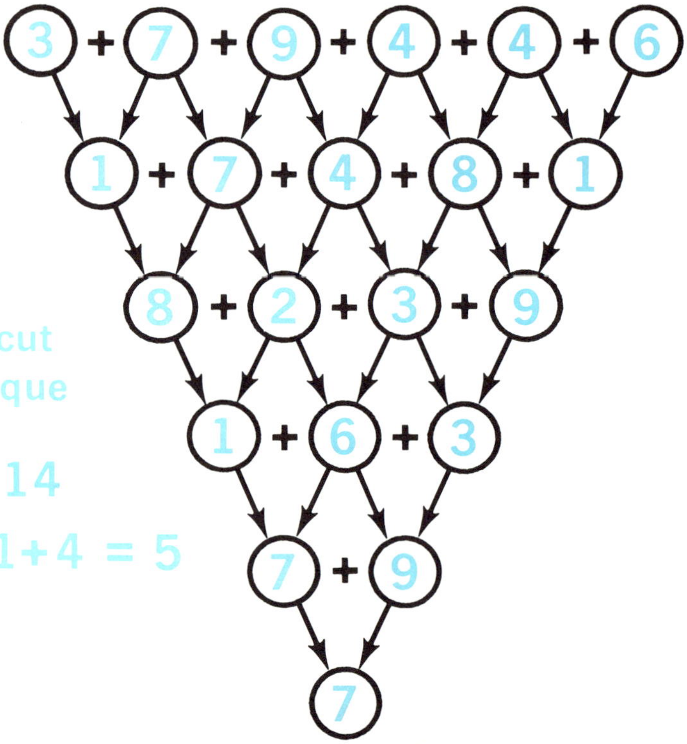

Approximately 90 mental calculations with only 4 answers to check.

CHECKING ADDITION

By Casting Out Nines

1)

495726	6
374815	1
152723	2
805142	2
614877	6
296351	8
2739634	7

2)

694012	4
408851	8
151973	8
347518	1
589362	6
291504	3
2483220	3

3)

258903	9
663751	1
163602	9
725859	9
250642	1
486576	9
2579333 (* under the 7)	2

* These are two deliberate mistakes in addition.

4)

$127.25	8
54.73	1
536.00	5
4.68	9
83.12	5
229.00	4
$1,034.78	5

5)

$ 43.10	8
681.53	5
2.75	5
35.50	4
328.00	4
96.76	1
$1,157.64 (* under the 5)	9

6)

$ 6.47	8
29.55	3
8.95	4
731.39	5
405.42	6
73.87	7
$1,255.65	6

Add and check the following problems.

7)

21.076	7
5.214	3
.058	4
72.500	5
9.029	2
51.008	5
158.885	8

Apex Number →

8)

583.04	2
79.00	7
155.85	6
43.00	7
64.42	7
300.08	2
1,225.39	4

9)

452.3	5
27.2	2
5841.4	4
128.0	2
200.0	2
94.9	4
6,743.8	1

NUMBER PATTERNS CHART

There are many number patterns hidden in this chart.
Use the letters to help identify their location.

A	B	C	D	E	F	G	H	I	J	
0	1	2	3	4	5	6	7	8	9	K
10	11	12	13	14	15	16	17	18	19	L
20	21	22	23	24	25	26	27	28	29	M
30	31	32	33	34	35	36	37	38	39	N
40	41	42	43	44	45	46	47	48	49	O
50	51	52	53	54	55	56	57	58	59	P
60	61	62	63	64	65	66	67	68	69	Q
70	71	72	73	74	75	76	77	78	79	R
80	81	82	83	84	85	86	87	88	89	S
90	91	92	93	94	95	96	97	98	99	T

Number Pattern Chart

NUMBER CHART

1	2	3	4	5	6	7	8	9	10	11	12	13	14	15
16	17	18	19	20	21	22	23	24	25	26	27	28	29	30
31	32	33	34	35	36	37	38	39	40	41	42	43	44	45
46	47	48	49	50	51	52	53	54	55	56	57	58	59	60
61	62	63	64	65	66	67	68	69	70	71	72	73	74	75
76	77	78	79	80	81	82	83	84	85	86	87	88	89	90
91	92	93	94	95	96	97	98	99	100	101	102	103	104	105
106	107	108	109	110	111	112	113	114	115	116	117	118	119	120
121	122	123	124	125	126	127	128	129	130	131	132	133	134	135
136	137	138	139	140	141	142	143	144	145	146	147	148	149	150
151	152	153	154	155	156	157	158	159	160	161	162	163	164	165
166	167	168	169	170	171	172	173	174	175	176	177	178	179	180
181	182	183	184	185	186	187	188	189	190	191	192	193	194	195
196	197	198	199	200	201	202	203	204	205	206	207	208	209	210
211	212	213	214	215	216	217	218	219	220	221	222	223	224	225

NUMBER CHART

(Title of Overlay)

1	2	3	4	5	6	7	8	9	10	11	12	13	14	15
16	17	18	19	20	21	22	23	24	25	26	27	28	29	30
31	32	33	34	35	36	37	38	39	40	41	42	43	44	45
46	47	48	49	50	51	52	53	54	55	56	57	58	59	60
61	62	63	64	65	66	67	68	69	70	71	72	73	74	75
76	77	78	79	80	81	82	83	84	85	86	87	88	89	90
91	92	93	94	95	96	97	98	99	100	101	102	103	104	105
106	107	108	109	110	111	112	113	114	115	116	117	118	119	120
121	122	123	124	125	126	127	128	129	130	131	132	133	134	135
136	137	138	139	140	141	142	143	144	145	146	147	148	149	150
151	152	153	154	155	156	157	158	159	160	161	162	163	164	165
166	167	168	169	170	171	172	173	174	175	176	177	178	179	180
181	182	183	184	185	186	187	188	189	190	191	192	193	194	195
196	197	198	199	200	201	202	203	204	205	206	207	208	209	210
211	212	213	214	215	216	217	218	219	220	221	222	223	224	225

Use overlay to color in different number patterns.
Place one overlay on another to find unique intersections.

Transparency Overlay

SQUARED NUMBERS

X	$(0)^2$	$(1)^2$	$(2)^2$	$(3)^2$	$(4)^2$	$(5)^2$	$(6)^2$	$(7)^2$	$(8)^2$	$(9)^2$
$(0)^2$	**0**	0	0	0	0	0	0	0	0	0
$(1)^2$	0	**1**	2	3	4	5	6	7	8	9
$(2)^2$	0	2	**4**	6	8	10	12	14	16	18
$(3)^2$	0	3	6	**9**	12	15	18	21	24	27
$(4)^2$	0	4	8	12	**16**	20	24	28	32	36
$(5)^2$	0	5	10	15	20	**25**	30	35	40	45
$(6)^2$	0	6	12	18	24	30	**36**	42	48	54
$(7)^2$	0	7	14	21	28	35	42	**49**	56	63
$(8)^2$	0	8	16	24	32	40	48	56	**64**	72
$(9)^2$	0	9	18	27	36	45	54	63	72	**81**

MULTIPLICATION TABLE

X	0	1	2	3	4	5	6	7	8	9
0	0	0	0	0	0	0	0	0	0	0
1	0	1	2	3	4	5	6	7	8	9
2	0	2	4	6	8	10	12	14	16	18
3	0	3	6	9	12	15	18	21	24	27
4	0	4	8	12	16	20	24	28	32	36
5	0	5	10	15	20	25	30	35	40	45
6	0	6	12	18	24	30	36	42	48	54
7	0	7	14	21	28	35	42	49	56	63
8	0	8	16	24	32	40	48	56	64	72
9	0	9	18	27	36	45	54	63	72	81

THE NUMBER LINE

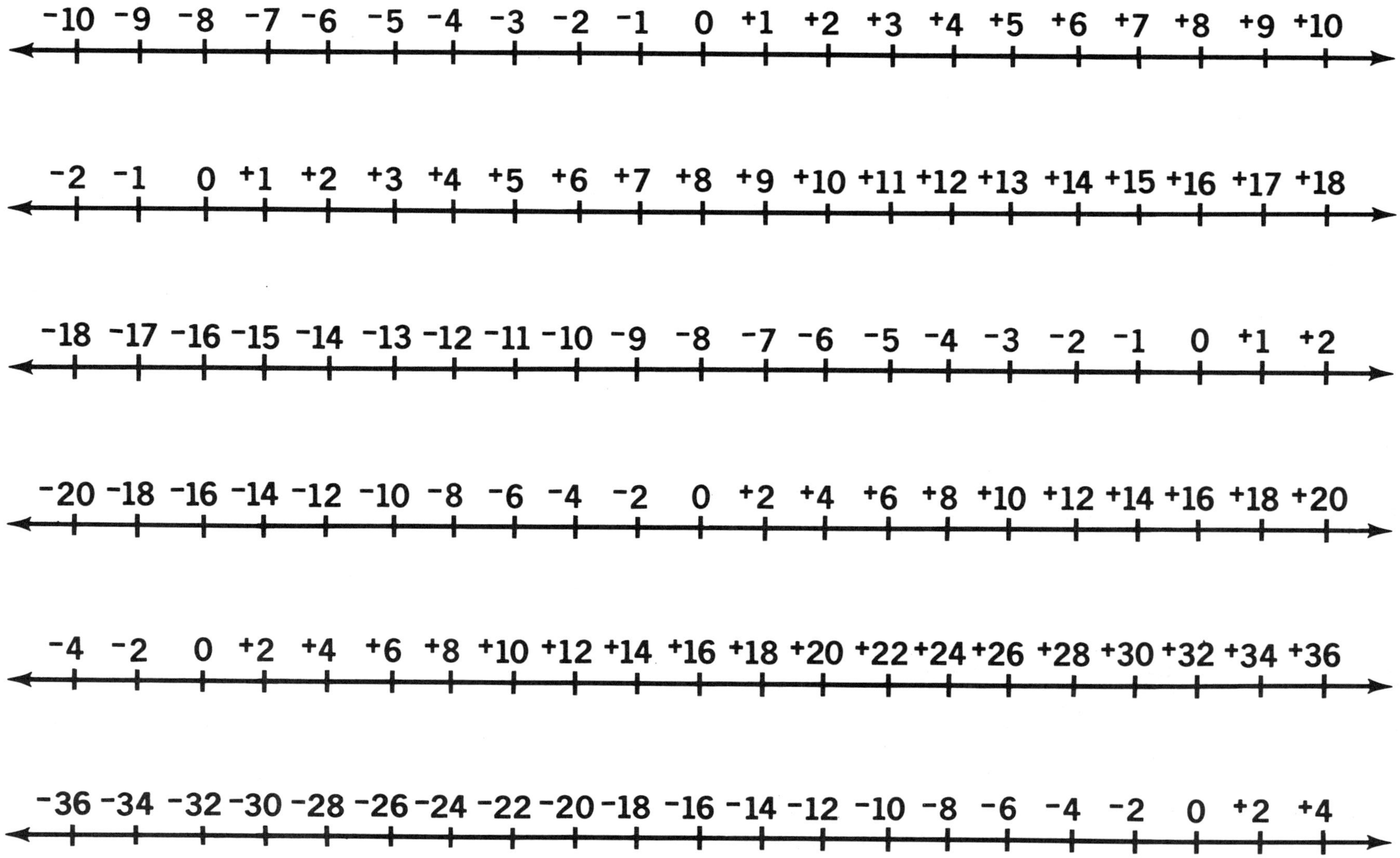

EXTRACTING SQUARE ROOT

PROBLEM —— $X^2 = 673.8$ OR $X = \sqrt{673.8}$

PROCEDURE ——

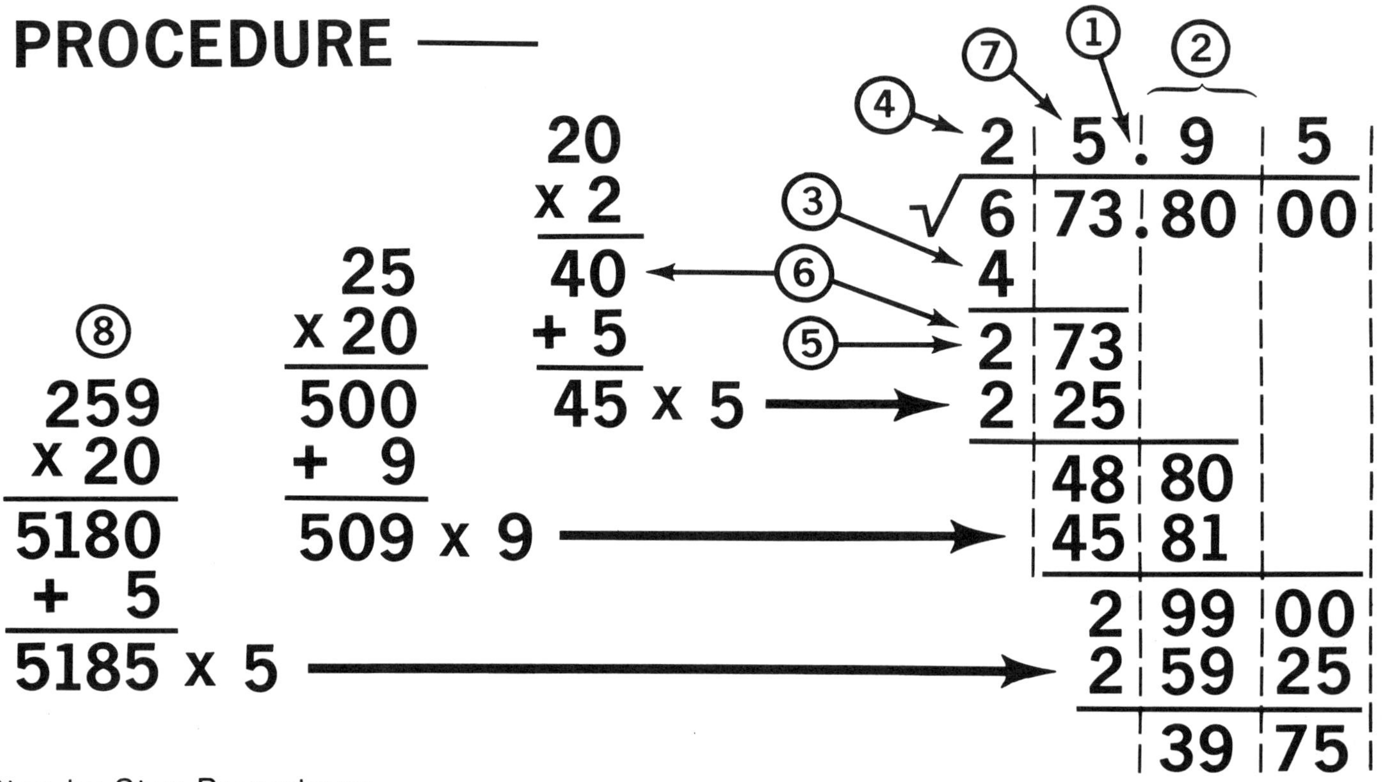

Step-by-Step Procedures

1. Place the decimal point directly over the decimal point in the number.
2. Beginning at the decimal point, separate the number into groups of two figures each, called Periods. Add zeros if needed.
3. Find the largest perfect square contained in the first period (6). The largest perfect square that can be subtracted from (6) is 4.
4. Write the 4 under the (6) and the square root of 4 (2) above the (6).
5. Subtract the perfect square 4 from the first period (6) and bring down the next period 73 to form the number 273.
6. Multiply the first number of the answer (2) by 20 to obtain a trial divisor of (40). Mentally divide the (40) into the remainder 273, to see about how many times it will go. It goes approximately 5 times.
7. Write the 5 above the next period to obtain an answer so far 25, then add the 5 to the 40 equalling 45. Multiply 45 by 5 and subtract the product from 273. The new dividend is 4880.
8. Repeat this process until you have gone past the last period.

SQUARES and SQUARE ROOTS

1	1	1.000 000
2	4	1.414 214
3	9	1.732 051
4	16	2.000 000
5	25	2.236 068
6	36	2.449 490
7	49	2.645 751
8	64	2.828 427
9	81	3.000 000
10	100	3.162 278
11	121	3.316 625
12	144	3.464 102
13	169	3.605 551
14	196	3.741 657
15	225	3.872 983
16	256	4.000 000
17	289	4.123 106
18	324	4.242 641
19	361	4.358 899
20	400	4.472 136
21	441	4.582 576
22	484	4.690 416
23	529	4.795 832
24	576	4.898 979
25	625	5.000 000

26	676	5.099 020
27	729	5.196 152
28	784	5.291 503
29	841	5.385 165
30	900	5.477 226
31	961	5.567 764
32	1 024	5.656 854
33	1 089	5.744 563
34	1 156	5.830 952
35	1 225	5.916 080
36	1 296	6.000 000
37	1 369	6.082 763
38	1 444	6.164 414
39	1 521	6.244 998
40	1 600	6.324 555
41	1 681	6.403 124
42	1 764	6.480 741
43	1 849	6.557 439
44	1 936	6.633 250
45	2 025	6.708 204
46	2 116	6.782 330
47	2 209	6.855 655
48	2 304	6.928 203
49	2 401	7.000 000
50	2 500	7.071 068

51	2 601	7.141 428
52	2 704	7.211 103
53	2 809	7.280 110
54	2 916	7.348 469
55	3 025	7.416 198
56	3 136	7.483 315
57	3 249	7.549 834
58	3 364	7.615 773
59	3 481	7.681 146
60	3 600	7.745 967
61	3 721	7.810 250
62	3 844	7.874 008
63	3 969	7.937 254
64	4 096	8.000 000
65	4 225	8.062 258
66	4 356	8.124 038
67	4 489	8.185 353
68	4 624	8.246 211
69	4 761	8.306 624
70	4 900	8.366 600
71	5 041	8.426 150
72	5 184	8.485 281
73	5 329	8.544 004
74	5 476	8.602 325
75	5 625	8.660 254

SQUARES and SQUARE ROOTS

76	5 776	8.717 798
77	5 929	8.774 964
78	6 084	8.831 761
79	6 241	8.888 194
80	6 400	8.944 272
81	6 561	9.000 000
82	6 724	9.055 385
83	6 889	9.110 434
84	7 056	9.165 151
85	7 225	9.219 544
86	7 396	9.273 618
87	7 569	9.327 379
88	7 744	9.380 832
89	7 921	9.433 981
90	8 100	9.486 833
91	8 281	9.539 392
92	8 464	9.591 663
93	8 649	9.643 651
94	8 836	9.695 360
95	9 025	9.746 794
96	9 216	9.797 959
97	9 409	9.848 858
98	9 604	9.899 495
99	9 801	9.949 874
100	10 000	10.000 000

FIBONACCI NUMBERS

1	1	121 393	26
1	2	196 418	27
2	3	317 811	28
3	4	514 229	29
5	5	832 040	30
8	6	1 346 269	31
13	7	2 178 309	32
21	8	3 524 578	33
34	9	5 702 887	34
55	10	9 227 465	35
89	11	14 930 352	36
144	12	24 157 817	37
233	13	39 088 169	38
377	14	63 245 986	39
610	15	102 334 155	40
987	16	165 580 141	41
1 597	17	267 914 296	42
2 584	18	433 494 437	43
4 181	19	701 408 733	44
6 765	20	1 134 903 170	45
10 946	21	1 836 311 903	46
17 711	22	2 971 215 073	47
28 657	23	4 807 526 976	48
46 368	24	7 778 742 049	49
75 025	25	12 586 269 025	50

LUCAS NUMBERS

1	1	271 443	26
3	2	439 204	27
4	3	710 647	28
7	4	1 149 851	29
11	5	1 860 498	30
18	6	3 010 349	31
29	7	4 870 847	32
47	8	7 881 196	33
76	9	12 752 043	34
123	10	20 633 239	35
199	11	33 385 282	36
322	12	54 018 521	37
521	13	87 403 803	38
843	14	141 422 324	39
1 364	15	228 826 127	40
2 207	16	370 248 451	41
3 571	17	599 074 578	42
5 778	18	969 323 029	43
9 349	19	1 568 397 607	44
15 127	20	2 537 720 636	45
24 476	21	4 106 118 243	46
39 603	22	6 643 838 879	47
64 079	23	10 749 957 122	48
103 682	24	17 393 796 001	49
167 761	25	28 143 753 123	50

DECIMAL EQUIVALENTS OF FRACTIONS

Fraction	Decimal	Fraction	Decimal
1/64	.015625	33/64	.515625
1/32	.03125	17/32	.53125
3/64	.046875	35/64	.546875
1/16	.0625	9/16	.5625
5/64	.078125	37/64	.578125
3/32	.09375	19/32	.59375
7/64	.109375	39/64	.609375
1/8	.125	5/8	.625
9/64	.140625	41/64	.640625
5/32	.15625	21/32	.65625
11/64	.171875	43/64	.671875
3/16	.1875	11/16	.6875
13/64	.203125	45/64	.703125
7/32	.21875	23/32	.71875
15/64	.234375	47/64	.734375
1/4	.25	3/4	.75
17/64	.265625	49/64	.765625
9/32	.28125	25/32	.78125
19/64	.296875	51/64	.796875
5/16	.3125	13/16	.8125
21/64	.328125	53/64	.828125
11/32	.34375	27/32	.84375
23/64	.359375	55/64	.859375
3/8	.375	7/8	.875
25/64	.390625	57/64	.890625
13/32	.40625	29/32	.90625
27/64	.421875	59/64	.921875
7/16	.4375	15/16	.9375
29/64	.453125	61/64	.953125
15/32	.46875	31/32	.96875
31/64	.484375	63/64	.984375
1/2	.5	1	1.

BASE SYSTEMS

PLACE VALUE CHART

$(N)^5$ +	$(N)^4$ +	$(N)^3$ +	$(N)^2$ +	$(N)^1$ +	$(N)^0$	Base N
32	16	8	4	2	1	2
243	81	27	9	3	1	3
1,024	256	64	16	4	1	4
3,125	625	125	25	5	1	5
7,776	1,296	216	36	6	1	6
16,807	2,401	343	49	7	1	7
32,768	4,096	512	64	8	1	8
59,049	6,561	729	81	9	1	9
100,000	10,000	1,000	100	10	1	10

Place Value Chart

BASE SYSTEMS

PLACE VALUE CHART

$(N)^5$ +	$(N)^4$ +	$(N)^3$ +	$(N)^2$ +	$(N)^1$ +	$(N)^0$	Base N
32	16	8	4	2	1	2
243	81	27	9	3	1	3
1,024	256	64	16	4	1	4
3,125	625	125	25	5	1	5
7,776	1,296	216	36	6	1	6
16,807	2,401	343	49	7	1	7
32,768	4,096	512	64	8	1	8
59,049	6,561	729	81	9	1	9
100,000	10,000	1,000	100	10	1	10

Transparency Overlay

BASE POWERS

n	$(2)^n$	$(3)^n$	$(4)^n$	$(5)^n$	$(6)^n$	$(7)^n$	$(8)^n$	$(9)^n$	$(10)^n$
0	1	1	1	1	1	1	1	1	1
1	2	3	4	5	6	7	8	9	10
2	4	9	16	25	36	49	64	81	100
3	8	27	64	125	216	343	512	729	1000
4	16	81	256	625	1296	2401	4096	6561	10000
5	32	243	1024	3125	7776	16807	32768	59049	100,000
6	64	729	4096	15625	46656	117,649	262,144	531,441	1,000, -000
7	128	2187	16384	78125	279936	823543	2,097, -152	4,782, -969	10,000, -000
8	256	6561	65536	390625	1,679, -616	5,764, -801	16,777, -216	43,046, -721	100,00 -0,000
9	512	19683	262144	1,953, -125	10,077, -696	40,353, -607	134,21 -7,728	387,42 -0,489	1,000,0 -00,000
10	1024	59049	1,048, -576	9,765, -625	60,466, -176	282,47 -5,249	1,073,74 -1,824	3,486,7 -84,401	10,000,0 -00,000

BASE POWERS

n	$(2)^n$	$(3)^n$	$(4)^n$	$(5)^n$	$(6)^n$	$(7)^n$	$(8)^n$	$(9)^n$	$(10)^n$
0	1	1	1	1	1	1	1	1	1
1	2	3	4	5	6	7	8	9	10
2	4	9	16	25	36	49	64	81	100
3	8	27	64	125	216	343	512	729	1000
4	16	81	256	625	1296	2401	4096	6561	10000
5	32	243	1024	3125	7776	16807	32768	59049	100,000
6	64	729	4096	15625	46656	117,649	262,144	531,441	1,000, -000
7	128	2187	16384	78125	279936	823543	2,097, -152	4,782, -969	10,000, -000
8	256	6561	65536	390625	1,679, -616	5,764, -801	16,777, -216	43,046, -721	100,00 -0,000
9	512	19683	262144	1,953, -125	10,077, -696	40,353, -607	134,21 -7,728	387,42 -0,489	1,000,0 -00,000
10	1024	59049	1,048, -576	9,765, -625	60,466, -176	282,47 -5,249	1,073,74 -1,824	3,486,7 -84,401	10,000,0 -00,000

Base Powers Chart

ADDITION

$$\begin{array}{r} 23 \\ 46 \\ +\,72 \\ \hline 141 \end{array}$$

23, 46, 72 — Addends

141 ← Sum

$$18 + 43 + 12 = 73$$

18, 43, 12 — Addends

SUBTRACTION

$$\begin{array}{r} 254 \\ -\,132 \\ \hline 122 \end{array}$$

254 ← Minuend

132 ← Subtrahend

122 ← Difference

$$387 - 251 = 136$$

387 — Minuend; 251 — Subtrahend

MULTIPLICATION

$$\begin{array}{r} 23 \\ \times\,18 \\ \hline 184 \\ 23 \\ \hline 414 \end{array}$$

23, 18 — Factors

184, 23 — Partial Product

414 ← Product

$$23 \times 18 = 414$$

23, 18 — Factors

DIVISION

$$4\overline{)3.00} = .75$$

.75 ← Quotient

3.00 ← Dividend

4 ← Divisor

$$\frac{3}{4}$$

3 ← Numerator

— ← Divided by

4 ← Denominator

$$3 \div 4 = .75$$

ADDITION

$$\begin{array}{r} 23 \\ 46 \\ +\,72 \\ \hline 141 \end{array}$$

23, 46, 72 → Addends

141 ← Sum

$18 + 43 + 12 = 73$

18 + 43 + 12 → Addends

73 → Sum

SUBTRACTION

$$\begin{array}{r} 254 \\ -\,132 \\ \hline 122 \end{array}$$

254 ← Minuend

132 ← Subtrahend

122 ← Difference

$387 - 251 = 136$

387 → Minuend

251 → Subtrahend

136 → Difference

MULTIPLICATION

$$\begin{array}{r} 23 \\ \times 18 \\ \hline 184 \\ 23 \\ \hline 414 \end{array}$$

23, 18 → Factors

184, 23 → Partial Product

414 ← Product

$23 \times 18 = 414$

23 × 18 → Factors

414 → Product

DIVISION

$$4\overline{)3.00}\ \text{with quotient}\ .75$$

.75 ← Quotient

3.00 ← Dividend

4 ← Divisor

$$\frac{3}{4}$$

3 ← Numerator

— ← Divided by

4 ← Denominator

$3 \div 4 = .75$

Terminology Test

VENN DIAGRAM

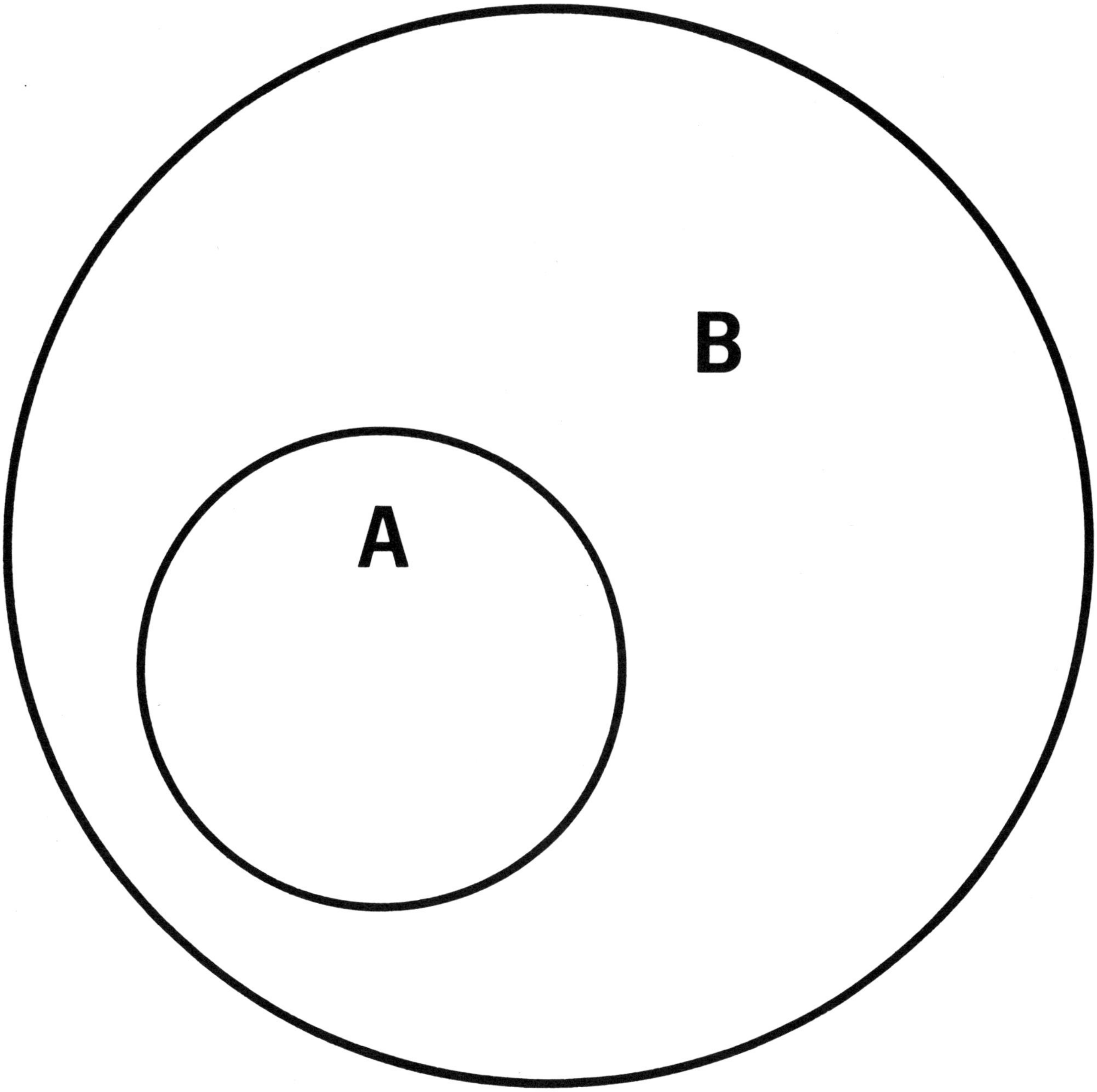

$A \subset B$

A is a subset of B

VENN DIAGRAM

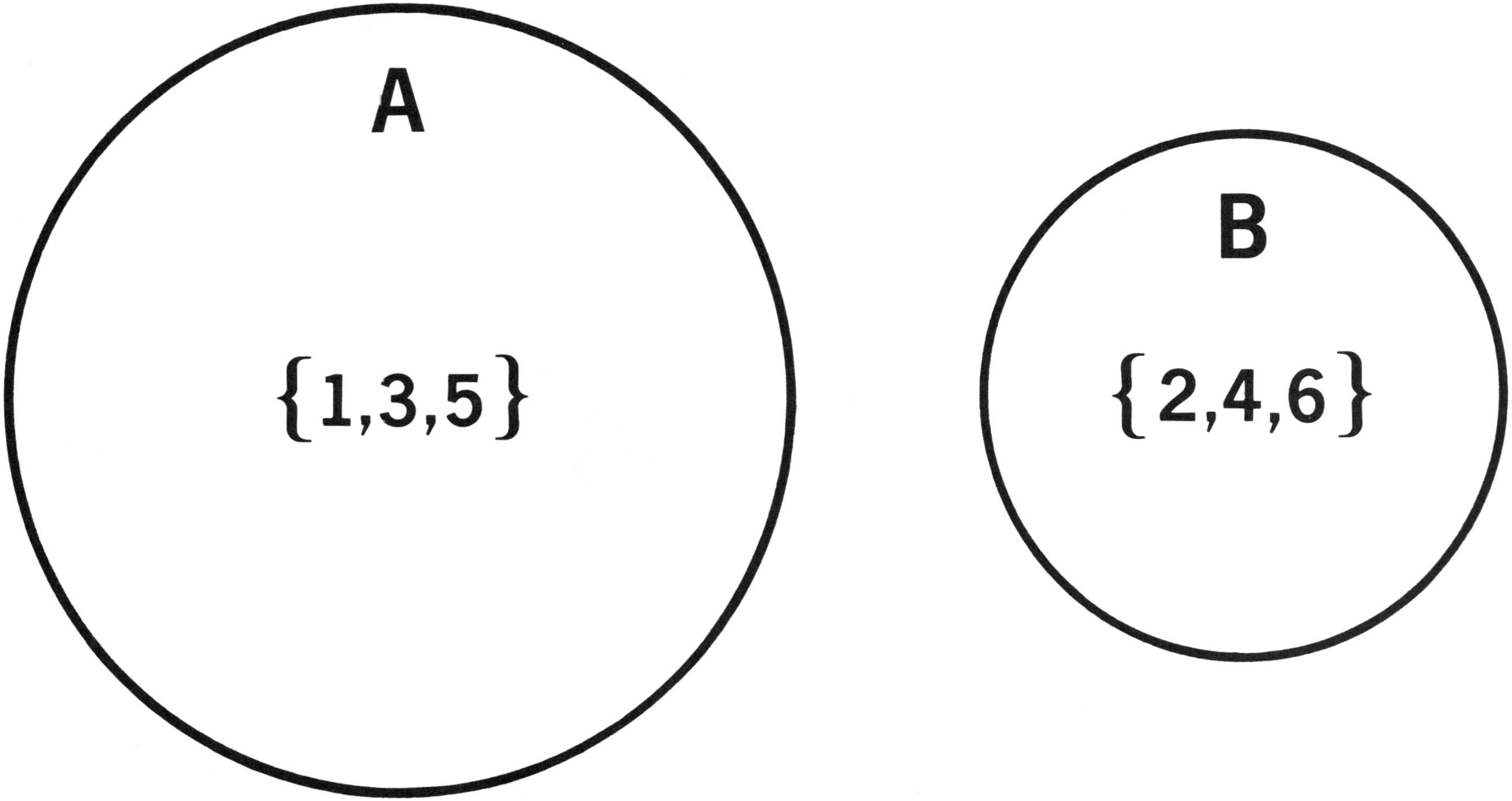

Sets A and B are Disjoint Sets

VENN DIAGRAM

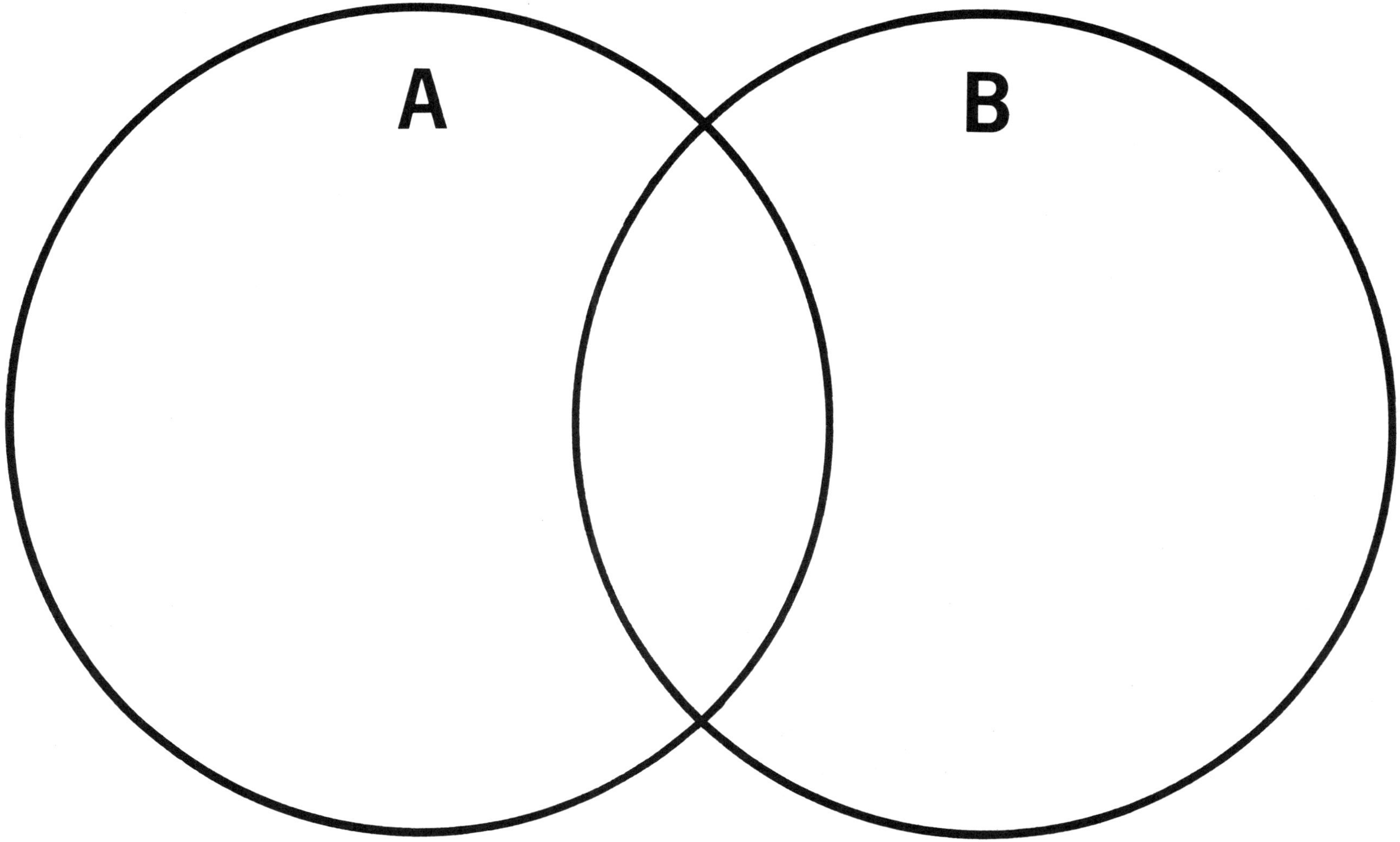

$A \cup B$

A union B

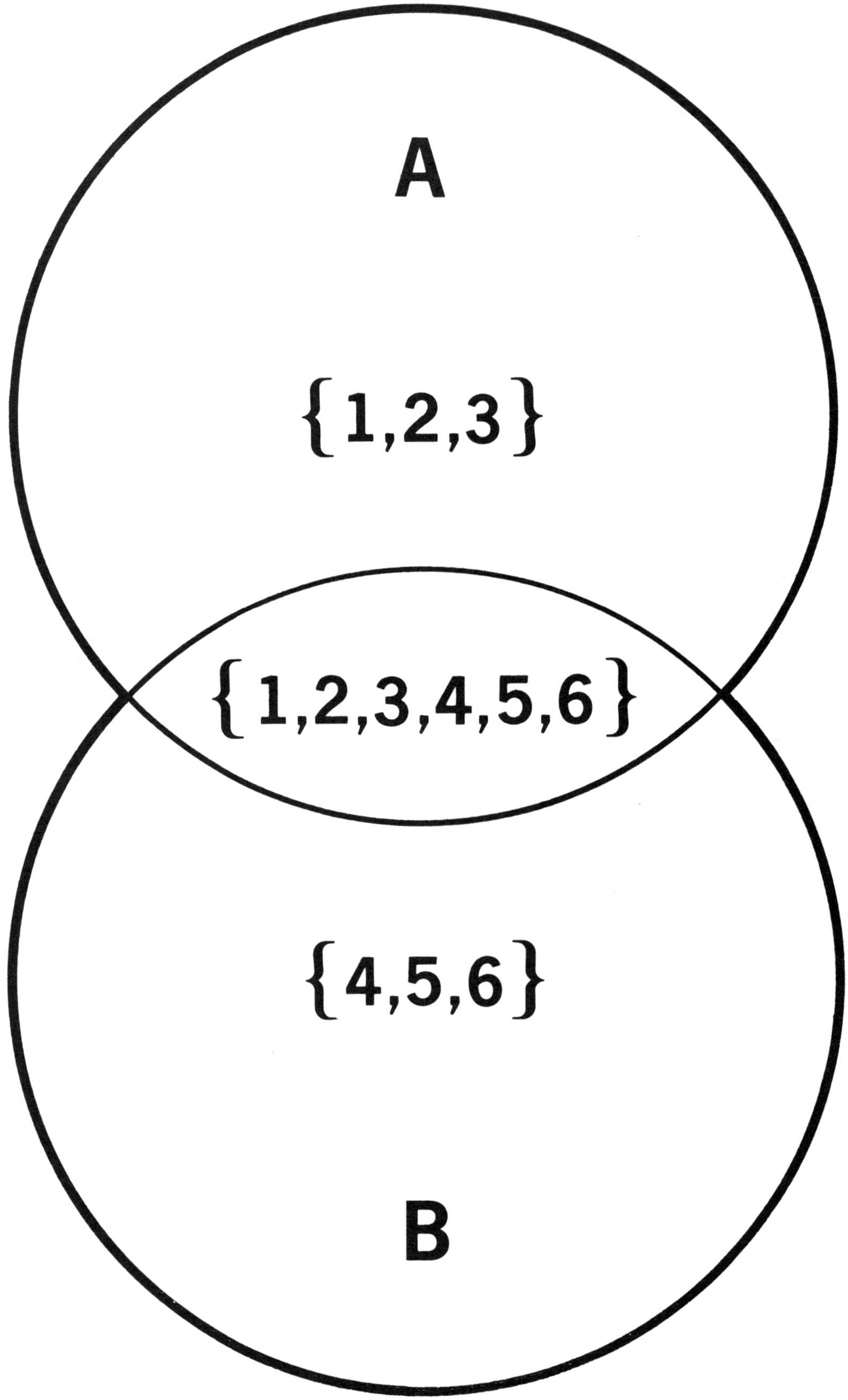
A
{1,2,3}
{1,2,3,4,5,6}
{4,5,6}
B

VENN DIAGRAM

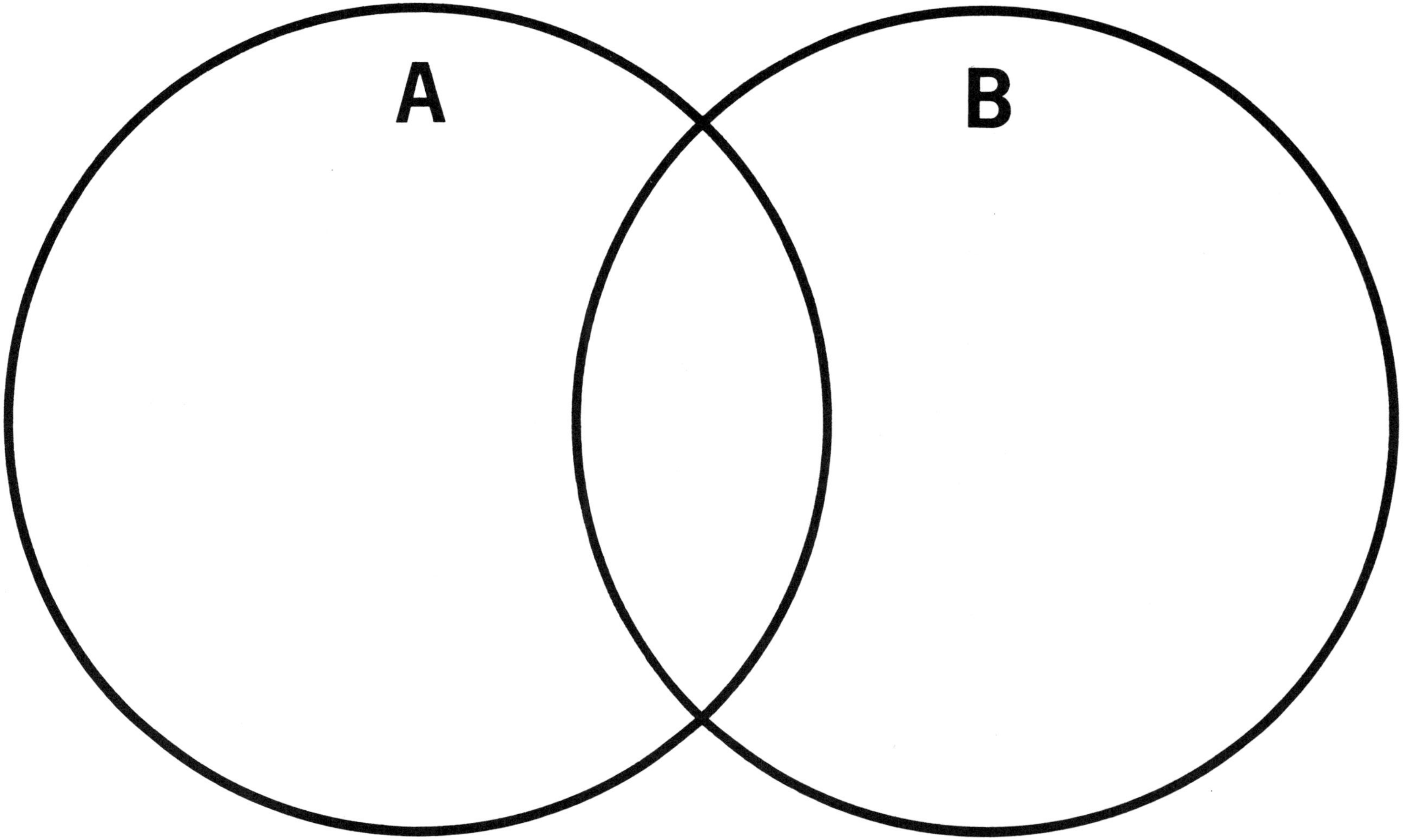

$A \cap B$

A intersect B

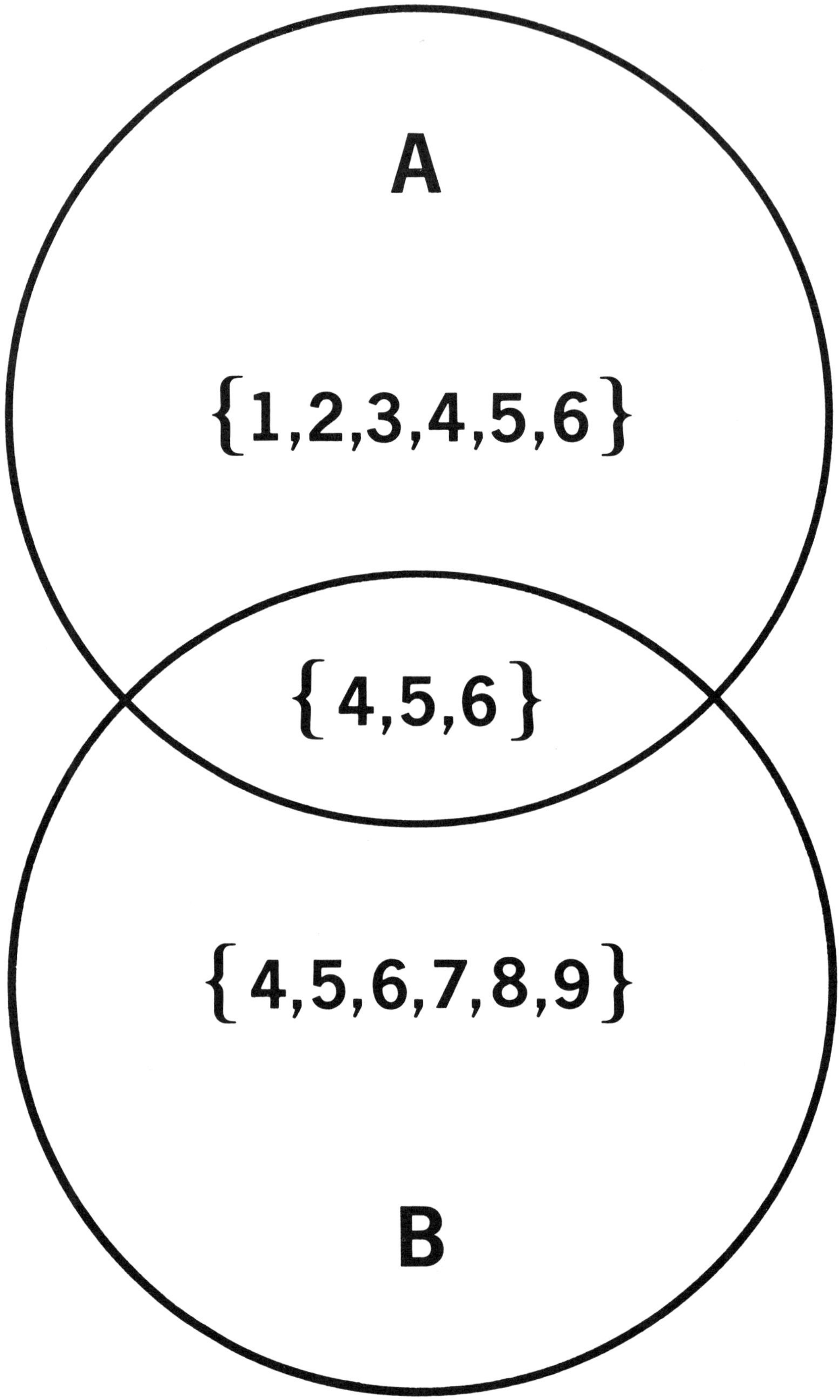
A
{1,2,3,4,5,6}
{4,5,6}
{4,5,6,7,8,9}
B

VENN DIAGRAM

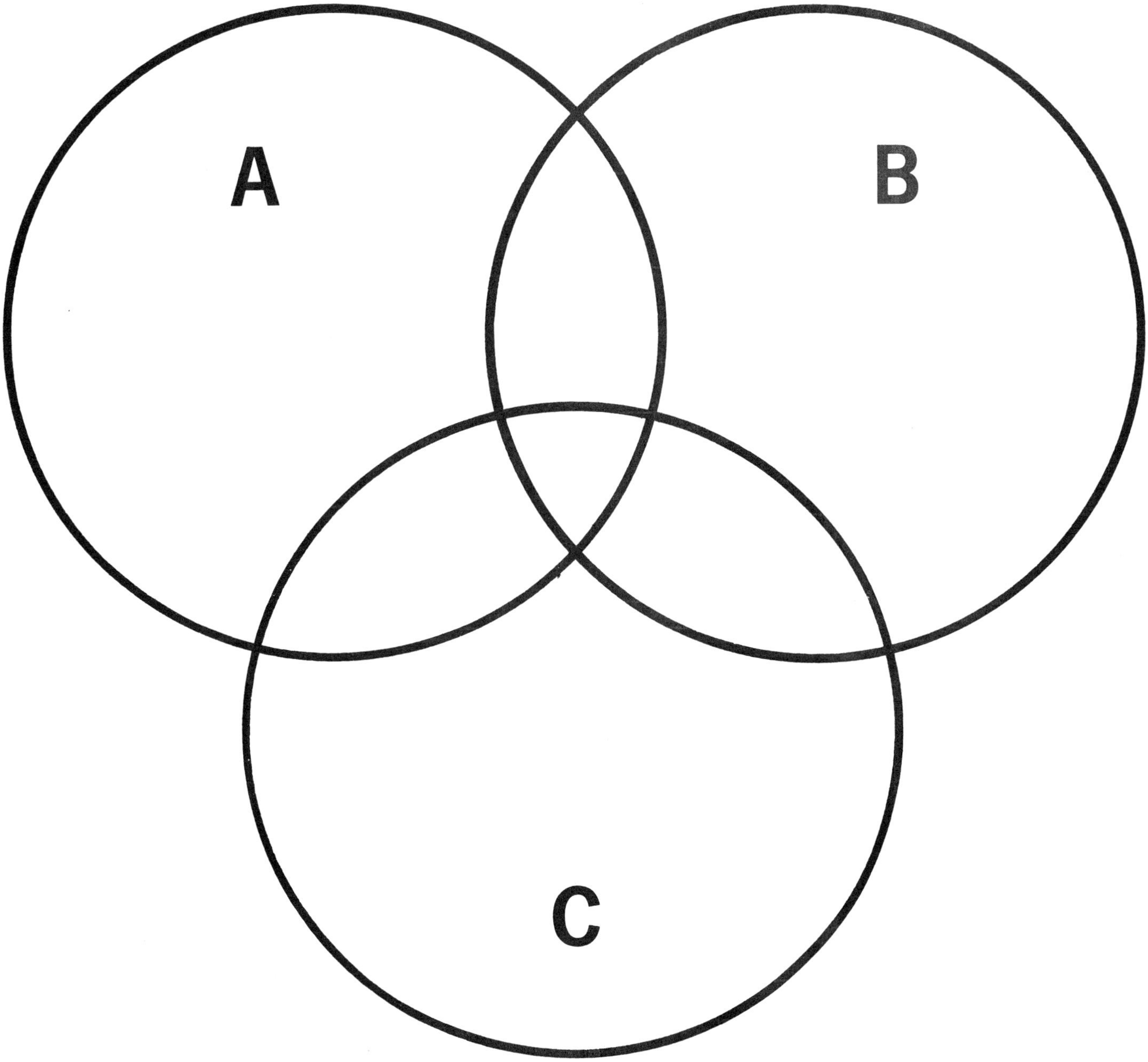

$(A \cap B) \cap C$

VENN DIAGRAM

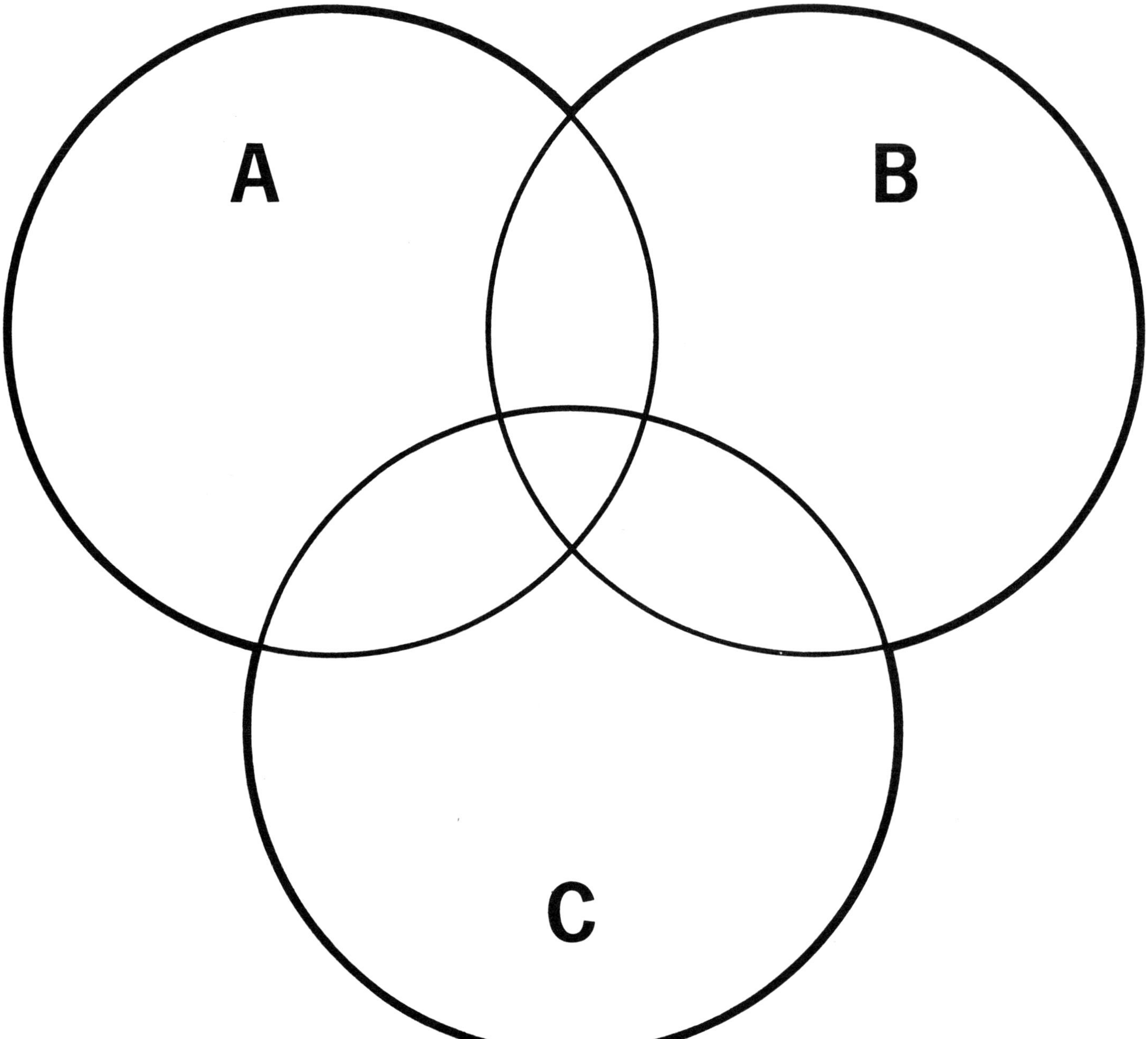

(A U B) U C

Date Last Name Score

Examination

1 ______ 11 ______
2 ______ 12 ______
3 ______ 13 ______
4 ______ 14 ______
5 ______ 15 ______
6 ______ 16 ______
7 ______ 17 ______
8 ______ 18 ______
9 ______ 19 ______
10 ______ 20 ______

Date Last Name Score

Examination

1 ______ 11 ______
2 ______ 12 ______
3 ______ 13 ______
4 ______ 14 ______
5 ______ 15 ______
6 ______ 16 ______
7 ______ 17 ______
8 ______ 18 ______
9 ______ 19 ______
10 ______ 20 ______

Date Last Name Score

Examination

1		13	
2		14	
3		15	
4		16	
5		17	
6		18	
7		19	
8		20	
9		21	
10		22	
11		23	
12		24	
		25	

Date Last Name Score

Square Numbers

Examination

1	1	13	169
2	4	14	196
3	9	15	225
4	16	16	256
5	25	17	289
6	36	18	324
7	49	19	361
8	64	20	400
9	81	21	441
10	100	22	484
11	121	23	529
12	144	24	576
		25	625

Date Last Name Score

Abundant Numbers

Examination

1	12	17	80
2	18	18	84
3	20	19	88
4	24	20	90
5	30	21	96
6	36	22	100
7	40	23	102
8	42	24	104
9	48	25	108
10	54	26	112
11	56	27	114
12	60	28	120
13	66	29	126
14	70	30	132
15	72	31	138
16	78	32	140
		33	144

Hexagon Number

Date Last Name Score

Pattern

Examination

1	1	17	817
2	7	18	919
3	19	19	1027
4	37	20	1141
5	61	21	1261
6	91	22	1387
7	127	23	1519
8	169	24	1657
9	217	25	1801
10	271	26	1951
11	331	27	2107
12	397	28	2269
13	469	29	2437
14	547	30	2611
15	631	31	2791
16	721	32	2977
		33	3169

Date | Last Name | Score

Sum of Squares

Examination

1	1	21	3311
2	5	22	3795
3	14	23	4324
4	30	24	4900
5	55	25	5525
6	91	26	6201
7	140	27	6930
8	204	28	7714
9	285	29	8555
10	385	30	9455
11	506	31	10416
12	650	32	11440
13	819	33	12529
14	1015	34	13685
15	1240	35	14910
16	1496	36	16206
17	1785	37	17575
18	2109	38	19019
19	2470	39	20540
20	2870	40	22140

Date | Last Name | Score

Circle Sections

Examination

1	2	21	232
2	4	22	254
3	7	23	277
4	11	24	301
5	16	25	326
6	22	26	352
7	29	27	379
8	37	28	407
9	46	29	436
10	56	30	466
11	67	31	497
12	79	32	529
13	92	33	562
14	106	34	596
15	121	35	631
16	137	36	667
17	154	37	704
18	172	38	742
19	191	39	781
20	211	40	821

Date	Last Name	Score

Triangular Numbers

Examination

1	1	26	351
2	3	27	378
3	6	28	406
4	10	29	435
5	15	30	465
6	21	31	496
7	28	32	528
8	36	33	561
9	45	34	595
10	55	35	630
11	66	36	666
12	78	37	703
13	91	38	741
14	105	39	780
15	120	40	820
16	136	41	861
17	153	42	903
18	171	43	946
19	190	44	990
20	210	45	1035
21	231	46	1081
22	253	47	1128
23	276	48	1176
24	300	49	1225
25	325	50	1275

Date	Last Name	Score

Pentagonal Numbers

Examination

1	5	26	1080
2	12	27	1162
3	22	28	1247
4	35	29	1335
5	51	30	1426
6	70	31	1520
7	92	32	1617
8	117	33	1717
9	145	34	1820
10	176	35	1926
11	210	36	2035
12	247	37	2147
13	287	38	2262
14	330	39	2380
15	376	40	2501
16	425	41	2625
17	477	42	2752
18	532	43	2882
19	590	44	3015
20	651	45	3151
21	715	46	3290
22	782	47	3432
23	852	48	3577
24	925	49	3725
25	1001	50	3876

Date Last Name Score

Spelling List

Examination

1 PRIME
2 UNION
3 UNIQUE
4 FACTOR
5 SUBSET
6 NUMBER
7 INVERSE
8 RADICAL
9 DECIMAL
10 NUMERAL
11 MULTIPLE
12 FRACTION
13 SOLUTION
14 EXPONENT
15 COMPOSITE
16 SYMMETRIC
17 RECIPROCAL
18 CONJECTURE
19 DENOMINATOR
20 ERATOSTHENES

Date Last Name Score

Spelling List

Examination

1 RADIUS
2 SQUARE
3 DEGREE
4 OBTUSE
5 INTEGER
6 OBLIQUE
7 SCALENE
8 AVERAGE
9 MEASURE
10 VERTICAL
11 DIAMETER
12 PRINCIPAL
13 ISOSCELES
14 PERIMETER
15 RECTANGLE
16 NUMERATOR
17 EQUILATERAL
18 COMMUTATIVE
19 CANCELLATION
20 DEPRECIATION

Date Last Name Score

Examination

1 ________	21 ________
2 ________	22 ________
3 ________	23 ________
4 ________	24 ________
5 ________	25 ________
6 ________	26 ________
7 ________	27 ________
8 ________	28 ________
9 ________	29 ________
10 ________	30 ________
11 ________	31 ________
12 ________	32 ________
13 ________	33 ________
14 ________	34 ________
15 ________	35 ________
16 ________	36 ________
17 ________	37 ________
18 ________	38 ________
19 ________	39 ________
20 ________	40 ________

Bibliography

AMICABLE NUMBERS—

Barnet, I. A., "The Fascination of Whole Numbers," *Arithmetic Teacher*, vol. 64, no. 2 (February, 1971), p.107.

Freitag, Herta and Arthur, *The Number Story*, National Council of Teachers of Mathematics, 1960, p. 50.

Larsen, Harold D., *This and That*, New York: Harper & Row, Publishers, 1963, pp. 12-13.

CASTING OUT NINES—

Johnson, Donovan A. and Glenn, William H., *Short Cuts in Computing*. Webster Publishing Company, 1961, pp. 39-42.

CIRCLE SECTIONS—

Johnson, Donovan A. and Glenn, William H., *Invitation to Mathematics*, Webster Publishing Company, 1960, pp. 7-8.

FACTORIAL PHENOMENA—

Madachy, Joseph S., "Some New Narcissistic Numbers," *The Fibonacci Quarterly*, April 1972, vol. 10, no. 3, pp. 295-298.

FIBONACCI AND LUCAS NUMBERS—

Adler, Irving, *Mathematics*, Golden Library of Knowledge, Golden Press, New York, 1961, pp. 16-17.

Alfred, Brother U., *An Introduction to Fibonacci Discovery*. The Fibonacci Association, San Jose State College, San Jose, California.

Curl, James C., "Fibonacci Numbers and the Slow Learner," *The Fibonacci Quarterly*, October, 1968.

Gardner, Martin, "Mathematical Games," *Scientific American*, March, 1969, pp.116-120.

Hogart, Verner E., *Fibonacci and Lucas Numbers*, Houghton Mifflin Company, 1969, pp. 6-56.

FIGURATE NUMBERS—

Freitag, Herta and Arthur, *The Number Story*, National Council of Teachers of Mathematics, 1960, p. 51.

Graflund, Robert S., *Patterns and Puzzles in Modern Math*, J. Weston Walch, 1970, pp. 18-19.

Smith, Joe K.,"The Nth Polygonal Number," *Mathematics Teacher*, vol. 65, no. 3 (March, 1972), pp. 221-225.

GAUSS' METHOD OF SUMMATION—

Denholm, Richard A., *Mathematics: Man's Key to Progress*, California State Department of Education, pp. 86-87.

Johnson, Donovan A. and Glenn, William H., *Invitation to Mathematics*, Webster Publishing Company, 1960, p. 22.

NARCISSISTIC NUMBERS—

Madachy, Joseph S., "Some New Narcissistic Numbers," *The Fibonacci Quarterly*, April 1972, vol. 10, no. 3, pp. 295-298.

PALINDROMIC NUMBERS—

Gardner, Martin, "Mathematical Games," *Scientific American*, August, 1970, pp. 110-112.

PASCAL'S TRIANGLE—

Gardner, Martin, "Mathematical Games," *Scientific American*, December, 1966, pp. 128-131.

Seymour, Dale, *Aftermath I*, Creative Publications, Inc., 1970, pp. 84-86.

PERFECT NUMBERS—

Molling, Joe, "Perfect Numbers," *The Mathematics Student Journal*, January, 1971, vol. 18, no. 2.

Reid, Constance, "Perfect Numbers," Prime Numbers and Perfect Numbers by School Mathematics Study Group. (Reprint Series, 1966)

PRIME NUMBERS—

Gardner, Martin, "Mathematical Games," *Scientific American*, March, 1969, pp. 116-120.

Larsen, Harold D., *Prime Numbers*, Harper & Row, Publishers, Inc., 1963.

More Notable Numbers

More Notable Numbers